Initial Value Methods
for Boundary Value Problems

This is Volume 100 in
MATHEMATICS IN SCIENCE AND ENGINEERING
A series of monographs and textbooks
Edited by RICHARD BELLMAN, *University of Southern California*

The complete listing of books in this series is available from the Publisher
upon request.

Initial Value Methods
for Boundary Value Problems

Theory and Application of Invariant Imbedding

GUNTER H. MEYER

School of Mathematics
Georgia Institute of Technology
Atlanta, Georgia

ACADEMIC PRESS   1973   New York and London

A Subsidiary of Harcourt Brace Jovanovich, Publishers

ACADEMIC PRESS, INC.
111 Fifth Avenue, New York, New York 10003

United Kingdom Edition published by
ACADEMIC PRESS, INC. (LONDON) LTD.
24/28 Oval Road, London NW1

LIBRARY OF CONGRESS CATALOG CARD NUMBER: 72-82626

AMS (MOS) 1970 Subject Classifications: 34B05, 34B10,
34B15, 34G05, 65L10

PRINTED IN THE UNITED STATES OF AMERICA

Meinen Eltern gewidmet

Contents

Chapter 3. Interface Problems

Preface

In this monograph we describe an initial value method for the solution of boundary value problems for ordinary differential equations. The resulting algorithm (but not its interpretation) is identical to the method of invariant imbedding so that this book fits into a field of applied mathematics which has seen active development in the last fifteen years. However, no systematic presentation of invariant imbedding as a numerical and analytical tool for the solution of boundary value problems has appeared in book form heretofore. We hope that this monograph will contribute toward filling this void.

Invariant imbedding is primarily a numerical method for boundary value problems. Such problems are rarely amenable to solution by one and only one method, but different methods will vary in their ease of application, speed of computation, and power of resolution. Invariant imbedding merits consideration as an additional, and essentially different, tool for boundary value problems. During the preparation of this material we have learned much about the practical aspects of the method. We have found that it is a fast, flexible, mechanically applicable, yet mathematically sound complement to the better known solution techniques such as shooting (also an initial value method), finite difference, or integral equation methods.

Our presentation is not formal. Instead, we try to show the geometric relationship between the boundary value problem and the invariant imbedding equations. This has enabled us to apply invariant imbedding to nonstandard problems such as free boundary value problems, and it may help in the future in modifying the method for problems not considered here. The geometric interpretation of invariant imbedding becomes possible because we imbed the given problem into a family of initial value problems rather than boundary value problems as is usually the case in the literature. Initial value problems for ordinary differential equations are much better understood than boundary value problems.

Except for Chapters 5 and 6, our presentation is elementary. This is not because we have restricted ourselves to simple boundary value problems but because the basic theory itself is simple. Indeed, we have tried to demonstrate the applicability of invariant imbedding to technical problems rather than rigged textbook examples. The reader need not be a mathematician. For an understanding of the first four chapters little more is required than acceptance of the basic existence theorem for the solution of ordinary differential equations (which for completeness is stated in the introduction) and a willingness to become familiar with our notation (also explained in the introduction). Chapters 5 and especially 6 are more demanding and need familiarity with the language of functional analysis.

Finally, while theory and application are not clearly separated, we have tried to present our results in such a manner that the reader interested in the practical aspects of invariant imbedding can skip the theoretical topics without penalty. In particular, we have tried to make Sections 2.4–4.2 self-contained. They concentrate on invariant imbedding as a numerical method for various fixed and free boundary value problems.

Acknowlegments

Most of the research and writing for this monograph were carried out at the Field Research Laboratory of Mobil Research and Development Corporation in Dallas, Texas, and I am grateful to my former employer for the opportunity to work on this book. I am indebted to Drs. James S. McNiel and Franz S. Selig for welcoming the project and making available the resources of FRL, and to Dr. Manus R. Foster whose strong support allowed me to write the monograph I had envisioned from the outset. I also would like to thank Mrs. Lynda Beaver for transcribing near illegible notes into typed copy, and Mrs. Sarah Brister for competently and cheerfully going through seemingly endless cycles of additions, deletions, and retypings.

Finally, I am grateful to my colleagues John R. Cannon and James V. Herod for reading and commenting on portions of the manuscript, and to the Georgia Institute of Technology for the opportunity to complete this work.

Initial Value Methods
for Boundary Value Problems

Introduction

0.1. AN OUTLINE OF THE MONOGRAPH

We shall present a brief survey over the material in this book, which will amplify the table of contents and may steer the reader toward those sections of interest and use to him.

Chapter 1 contains the background material necessary for the conversion of boundary value into initial value problems. Our vehicle for this conversion is the classical theory of characteristics for partial differential equations with the same principal part. In Section 1.1 the terminology and geometric concepts are introduced and it is shown qualitatively and quantitatively how ordinary differential equations, when interpreted as characteristic equations, can piece together an integral surface for an associated partial differential equation. In Section 1.2 we restrict ourselves to the case of linear inhomogeneous characteristic differential equations. It is shown that the characteristics are now related through an "affine" transformation (a generalized Riccati transformation), which can be computed from a matrix Riccati differential equation. The properties of the Riccati equation are studied in Section 1.3 because it is intimately connected with the solution of linear boundary value problems. In this context special emphasis is placed on Hermitian Riccati equations which are more tractable and which arise naturally in optimal control problems. Section 1.4 contains some comments on the solvability of partial differential equations with the same principal part when the characteristic equations are nonlinear ordinary differential equations. The mostly negative conclusions motivate the search for numerical methods taken up in the next chapter. We conclude the discussion of characteristic theory with a brief and rather formal treatment of the case where the characteristic equations are difference equations.

Chapter 2 is the heart of the monograph where the concepts of characteristic theory are applied to solve two-point boundary value problems for ordinary differential equations. To place our approach into a historical and mathematical perspective, we present in Section 2.1 a survey over various initial value techniques proposed for boundary value problems. Important for our theory is the explicit shooting method which is briefly stated. In Section 2.2 the differential equations of the boundary value problem are interpreted as characteristic equations, and an initial value problem is derived for the associated partial differential equation with the same principal part through an implicit shooting method. This partial differential equation is the so-called invariant imbedding equation which is our basic tool for the solution of boundary value problems by initial value methods. We use in Section 2.3 the invariant imbedding approach and the quantitative results of characteristic theory to obtain existence and uniqueness theorems for two-point boundary value problems with linear, nonlinear and coupled boundary conditions. The main value of invariant imbedding, however, lies in its applicability to the numerical solution of boundary value problems. This topic is discussed at length in Section 2.4, where the solution algorithm is explicitly stated and where several finite difference methods for the integration of the partial differential invariant imbedding equation are described. The numerical examples of Section 2.5 are meant to illustrate and reinforce the comments of the preceding section on invariant imbedding for nonlinear equations. In Section 2.6, linear boundary value problems are considered for which invariant imbedding is particularly well adapted because the affine Riccati transformation reduces the partial differential equation to a system of ordinary differential equations. The numerical solution of these invariant imbedding equations is discussed in detail in Section 2.7 which also contains error bounds. Some of the points brought out in this discussion are taken up again in the numerical examples of Section 2.8. In the next section we briefly treat the case of linear inhomogeneous equations subject to one nonlinear boundary condition. Sufficient geometric insight has now been gained to tackle the class of free boundary value problems which arise frequently in practice but which are rarely treated in books on boundary value methods. We shall outline a rigorous solution algorithm and demonstrate the applicability of the method by solving some representative problems from optimal control and heat transfer. Finally, in Section 2.11, the discrete characteristic theory of 1.5 is applied to two-point boundary value problems for difference equations.

Chapter 3 takes the theory a step further by introducing interface (discontinuity) conditions in the interval of integration. The basic theory is given in 3.1, and in an example the permissible discontinuities in the differential equation are described. Section 3.2 notes the simplification for linear problems through use of the affine transformation. We then pick up the topic of Section 2.10 again but allow the interface rather than the boundary to be part of the solution. As an illustration, free interface problems for a convective and a conductive system are solved. In Section 3.4 it is shown that the invariant imbedding equation is a special case of the dynamic programming equation. An application of the formalism to a linear regulator control problem is described. In Section 3.5 linear eigenvalue problems for systems of ordinary differential equations are considered. The method is illustrated by solving for the smallest eigenvalue of a three-dimensional second-order differential equation.

Chapter 4 describes the application of invariant imbedding to multipoint boundary value problems. The general theory, an existence theorem for multipoint problems, and a numerical example from boundary layer theory, are given in Section 4.1. We concentrate on linear problems in Section 4.2.

Chapter 5 develops the theory of invariant imbedding for abstract boundary value problems. The intent here is to provide the tools for deriving the imbedding equations for neutron transport processes which historically constitute the core of the method of invariant imbedding. Section 5.1 deals with infinite dimensional differential equations with smooth right-hand sides, while Section 5.2 concentrates on linear evolution equations with (possibly) nonsmooth right-hand sides. Three examples involving the Boltzmann transport equation illustrate the application of the initial value technique. Section 5.3 covers boundary value problems for linear differential equations with continuous but not necessarily differentiable solutions. The validity of a nondifferentiable Riccati transformation is established, which is then applied to a distributed control problem.

Chapter 6 deals with the solution of infinite-dimensional Riccati equations. Section 6.1 treats the question of existence and uniqueness of local mild and strong solutions in a Banach space. Section 6.2 contains conditions for the existence of global solutions for non-Hermitian equations with bounded coefficients, while Section 6.3 covers the case of unbounded coefficients. The last section considers bounded solutions for Hermitian equations with bounded and unbounded coefficients.

0.2. NOTATION

The first four chapters deal with equations defined on real intervals and with values in a k-dimensional Euclidean space E^k. A vector $u \in E^k$ is always thought to be a column vector although for ease of printing it is written as the transpose of a row vector, namely $u = (u_1, \ldots, u_k)^T$, where u_j is the jth component of u and where T indicates the transpose. The scalar product of two vectors $u, x \in E^k$ is written consistently in inner product form as $\langle u, x \rangle = \sum_{j=1}^k u_i x_i$. For the expression $\langle u, u \rangle$ we use the standard (l_2) norm notation and write $\langle u, u \rangle = \| u \|^2$.

Except for several examples where the physical model preempted time, we have chosen t as the independent real variable in our differential equations. Hence for a function $u(t)$ defined on some interval I with values in E^k we denote by $u'(t)$ or simply u' the derivative of u with respect to t, so that $u' = (u_1', \ldots, u_k')^T$. The derivative notation is somewhat more involved for functions mapping E^n into E^m. If such a function is written as $w = w(x)$ where $x \in E^n$ and $w \in E^m$, then we shall need the derivative (more precisely, the Frechet derivative) $w_x(x)$ at x. This expression is shorthand for the Jacobian matrix

$$w_x(x) = \begin{pmatrix} \partial w_1/\partial x_1 & \cdots & \partial w_1/\partial x_n \\ \vdots & & \vdots \\ \partial w_m/\partial x_1 & \cdots & \partial w_m/\partial x_n \end{pmatrix}$$

evaluated at the point x. If F is a function defined on the cartesian product $I \times E^m \times E^n$ with values in E^m then we can consider partial derivatives at (t, u, x). These are denoted as follows:

$F_t(t, u, x)$ = partial derivative with respect to the scalar parameter t;

$F_u(t, u, x)$ = square Jacobian matrix $(\partial F_i/\partial u_j)$, $i, j = 1, \ldots, m$;

$F_x(t, u, x)$ = Jacobian matrix $(\partial F_i/\partial x_j)$, $i = 1, \ldots, m$; $j = 1, \ldots, n$.

Occasionally it is convenient to use the synonymous notation

$$F_t(t, u, x) = \frac{\partial F}{\partial t}(t, u, x), \qquad F_u(t, u, x) = \frac{\partial F}{\partial u}(t, u, x),$$

$$F_x(t, u, x) = \frac{\partial F}{\partial x}(t, u, x)$$

If $E^m = E^n = E^1$, then, of course, we have the usual scalar partial derivatives. For $m, n > 1$ the partial derivatives with respect to u and x will always be Jacobian matrices.

Additional special notation is introduced as needed throughout the text, particularly in Section 1.3 on the Hermitian Riccati equation, in Chapter 4 on multipoint problems, and in Chapters 5 and 6 on abstract equations.

0.3. BACKGROUND MATERIAL

Throughout this monograph, a given boundary value problem is considered as good as solved once it has been reduced to an initial value problem for ordinary differential equations. This would appear to be an optimistic attitude since the analytical and numerical solution of initial value problems in general cannot be regarded as straightforward. However, we invariably require that the differential equations of the boundary value problem are so well behaved that initial value problems for them are readily handled with the basic existence and uniqueness theory for ordinary differential equations. For the sake of completeness we shall collect here those theorems which are used repeatedly throughout the text. A complete development of the general theory in a coordinate free setting may be found in Dieudonné (1969).

A function w defined on E^n and with values in E^m is said to be locally Lipschitz continuous on a set $D \subset E^n$ if for each $z \in D$ there exists a neighborhood $B \subset D$ of z for which there is a constant K such that $\| w(x) - w(y) \| \leq K \| x - y \|$ for $x, y \in B$. If K is independent of z, then w is Lipschitz continuous on D and if $D = E^n$ then w is uniformly Lipschitz continuous on E^n. It is readily seen that if w is differentiable on D, then w is locally Lipschitz continuous on D, and if the derivative is uniformly bounded, then w is Lipschitz continuous on all of D. If we deal with a function $F(t, y)$ defined on a real interval I and some set $D \subset E^n$, then F is (locally) Lipschitz continuous in y if it is (locally) Lipschitz continuous for each $t \in I$. With this terminology we can state the basic existence and uniqueness result for ordinary differential equations.

THEOREM. Let $F: I \times D \subset I \times E^m \to E^m$ be continuous in t and locally Lipschitz continuous in y; then the initial value problem

$$y' = F(t, y), \qquad y(t_0) = y_0$$

with $t_0 \in I$ and $y_0 \in D$ has a unique (local) solution $y(t, t_0, y_0)$ for all t in a neighborhood of t_0. If F is p times differentiable in y then $y(t, t_0, y_0)$ is p times differentiable with respect to y_0 in a neighborhood of y_0.

Since our boundary value problems will be formulated for a given interval $[0, T]$, we need to continue the local solution over the entire interval. The general theory states that $y(t, t_0, y_0)$ can be continued in t until it either blows up (in norm) or y reaches the boundary of D. In most of the applications we shall require that F be differentiable with respect to y on E^n and that the derivative $F_y(t, y)$ be uniformly bounded. In this case $y(t, t_0, y_0)$ will exist over $[0, T]$ and be differentiable with respect to y_0.

To determine in general whether a solution $y(t, t_0, y_0)$ can be continued over a given interval, and also to compare solutions of the same differential equation subject to different initial values, the so-called Gronwall inequality is useful. It may be stated as follows. Let u, v, w be continuous functions for $t \geq 0$ and suppose that $w \geq 0$. Then

$$u(t) \leq v(t) + \int_0^t w(s)u(s)\,ds$$

implies the inequality

$$u(t) \leq v(t) + \int_0^t w(s)v(s) \exp\left(\int_0^t w(r)\,dr\right) ds$$

Note that if v is monotonely increasing, then $v(s)$ may be replaced by $v(t)$ and taken out of the integral. Integration of the remaining expression leads to the weaker but simpler bound

$$u(t) \leq v(t) \exp\left(\int_0^t w(r)\,dr\right)$$

In most of the applications throughout the text the functions u, v, and w are identified with the norms of continuous vector-valued functions. Typically, we may have a two-dimensional vector $(\| x(t) \|, \| y(t) \|)^{\mathrm{T}}$ and set $u(t) = \max\{\| x(t) \|, \| y(t) \|\}$.

The problem of continuing and estimating the solution of a linear equation of the type

$$y' = A(t)y + F(t), \qquad y(t_0) = y_0$$

becomes simple because $y(t)$ can be expressed by the variation of constants formula

$$y(t) = \phi(t, t_0)y_0 + \int_{t_0}^t \phi(t, r)F(r)\,dr$$

Here $\phi(t, r)$ is the fundamental solution of the matrix equation

$$\phi(t, r) = A(t)\phi(t, r), \qquad \phi(r, r) = I$$

where I is the $n \times n$ identity matrix. Let us briefly touch on the possibility of estimating $y(t)$ over an interval. Since the matrix equation can also be written in Volterra integral equation form as

$$\phi(t, r) = I + \int_r^t A(s)\phi(s, r)\, ds$$

and since for any vector norm

$$\| \phi(t, r) \| \le 1 + \int_r^t \| A(s) \|\, \| \phi(s, r) \|\, ds$$

it follows from Gronwall's inequality that

$$\| \phi(t, r) \| \le \exp\left(\int_r^t \| A(r) \|\, dr \right)$$

Hence if $A(t)$ is bounded on an interval $[0, T]$ then $\phi(t, r)$ is bounded for all $t, r \in [0, T]$. This in turn implies that $y(t)$ will remain bounded on $[0, T]$ if F is integrable.

At various stages in the development of our theory we are faced with the solution of nonlinear vector-valued problems which we generally express as

$$g(x) = 0$$

Here g is always a continuously differentiable function defined on E^n and with values in E^n. In the context of this monograph, it has proved convenient to attack such problems with a homotopy method. Suppose that H is a mapping from $[0, 1] \times E^n$ into E^n which is continuous in both variables and which satisfies $H(0, x_0) = 0$ for some known vector x_0, and $H(1, x) \equiv g(x)$. It follows that if the imbedding equation (homotopy) $H(\lambda, x) = 0$ has a continuous solution $x(\lambda)$ for $\lambda \in [0, 1]$, then $x(1)$ is a root of $g(x) = 0$. As examples of useful homotopies we cite $H(\lambda, x) = (\lambda - 1)g(x_0) + g(x)$ and $H(\lambda, x) = (1 - \lambda)(x - x_0) + \lambda g(x)$ for arbitrary $x_0 \in E^n$. The mechanism of continuing $x(\lambda)$ over $[0, 1]$ from x_0 to $x(1)$ can be based on iterative or differential equations methods. We are not concerned about minimal smoothness conditions here and shall choose the continuation provided by the differential equation

$$\frac{dH}{d\lambda}(\lambda, x) = H_x(\lambda, x)\frac{dx}{d\lambda} + H_\lambda(\lambda, x) = 0$$

If H_x is nonsingular we obtain the differential equation

$$dx/d\lambda = H_x^{-1}(\lambda, x)H_\lambda(\lambda, x), \qquad x(0) = x_0$$

If H_λ and H_x^{-1} are continuous over $[0, 1] \times E^n$ then this equation has a solution $x(\lambda)$ for all $\lambda \in [0, 1]$, and hence $x(1)$ solves $g(x) = 0$. For the two homotopies cited above the differential equation assumes the form

$$dx/d\lambda = -g_x^{-1}(x)g(x_0)$$

and

$$dx/d\lambda = -[(1 - \lambda)I + \lambda g_x(x)]^{-1}[g(x) - x]$$

in both cases with the initial value $x(0) = x_0$. Note that the first differential equation assures that $g(x) - b = 0$ always has a unique solution if $g_x(x)$ has a uniformly bounded inverse on E^n. In this case g is a homeomorphism on E^n.

For the second differential equation, it frequently becomes necessary to establish the existence of the inverse for the operator $[I + \lambda(g_x(x) - I)]$. Banach's lemma (also called Neumann's lemma) is a crude but useful tool for this purpose. It may be stated as follows:

LEMMA. If the norm of the matrix B satisfies $\| B \| < 1$, then the matrix $(I - B)$ is invertible and satisfies

$$(I - B)^{-1} = \sum_{n=0}^{\infty} B^n$$

and

$$\| (I - B)^{-1} \| \le \sum_{n=0}^{\infty} \| B \|^n = 1/(1 - \| B \|)$$

For a complete discussion of homotopy theory in E^n we refer to Ortega and Rheinboldt (1970).

One more major theorem used repeatedly throughout the monograph is the implicit function theorem which can be stated as follows.

THEOREM [Apostol (1957)]. Let $F = (F_1, \ldots, F_n)$ be a vector-valued function defined on an open set S in $E^n \times E^k$ with values in E^n. Suppose that F is continuously differentiable on S. Let (x_0, y_0) be a point in S for which $F(x_0, y_0) = 0$ and for which the Jacobian matrix $F_x(x_0, y_0)$ is nonsingular. Then there exists a k-dimensional neighborhood D of y_0 and a unique differentiable function g from E^n to E^n such that $g(y_0) = x_0$ and $F(g(y), y) = 0$ for all $y \in D$.

Some additional background material may be found at the beginning of Chapter 5. It is not needed for the development of invariant imbedding in finite-dimensional spaces.

Chapter

1

Characteristic Theory

1.1. FIRST-ORDER PARTIAL DIFFERENTIAL EQUATIONS
WITH THE SAME PRINCIPAL PART

All the material of this monograph is based on the relationship between certain first-order partial differential equations and their characteristic ordinary differential equations. There exists an extensive theory on this topic (including the Hamilton–Jacobi theory) which deals with single linear and nonlinear equations as well as certain systems of partial differential equations; for a complete exposition the reader is referred to standard texts on classical partial differential equations (see N.1.1). For the conversion of boundary into initial value problems we shall be concerned only with that small fraction of the general theory which deals with first-order partial differential equations with the same principal part. Such equations are actually hyperbolic partial differential equations but warrant separate consideration because of their special structure.

Consider the system of m partial differential equations

$$\sum_{j=1}^{n} (\partial u_i/\partial x_j)G_j(u, x) = F_i(u, x), \qquad i = 1,\ldots, m$$

or in vector form

$$u_x(x)G(u, x) = F(u, x) \qquad (1.1.1)$$

where $u = (u_1, \ldots, u_m)^{\mathrm{T}}$, $x = (x_1, \ldots, x_n)^{\mathrm{T}}$, $G = (G_1, \ldots, G_n)^{\mathrm{T}}$, and $F = (F_1, \ldots, F_m)^{\mathrm{T}}$. Because the same coefficients G_j appear in each equation, the system (1.1.1) is said to have the *same principal part*.

Assume that the vector-valued function $u = w(x)$ is a solution of Eq. (1.1.1). Then it follows that the normal n_i to the surface $\Phi(u_i, x) \equiv u_i - w_i(x) = 0$ in the $(n + 1)$-dimensional (u_i, x)-space is perpendicular to the vector $(F_i, G_1, \ldots, G_n)^{\mathrm{T}}$. Indeed, we have by definition of the normal

$$n_i = (1, -\partial w_i/\partial x_i, \ldots, -\partial w_i/\partial x_n)^{\mathrm{T}}$$

so that

$$\langle (F_i, G_1, \ldots, G_n)^{\mathrm{T}}, n_i \rangle \equiv \langle (F_i, G_1, \ldots, G_n)^{\mathrm{T}},$$

$$(1, -\partial w_i/\partial x_1, \ldots, -\partial w_i/\partial x_n)^{\mathrm{T}} \rangle = F_i - \sum_{j=1}^{n} (\partial w_i/\partial x_j) G_j = 0$$

for $i = 1, \ldots, m$. In other words, the direction field defined by

$$du_i/dr = F_i(u, x)$$
$$dx_j/dr = G_j(u, x), \qquad j = 1, \ldots, n \tag{1.1.2}$$

where r is a free parameter, is tangent to $w_i(x)$ at every point of $u = w(x)$. A solution of (1.1.1) is called an *integral surface*, Eqs. (1.1.2) for $i = 1, \ldots, m$, $j = 1, \ldots, n$ are known as the *characteristic equations* corresponding to (1.1.1), and their solution $\{u(r), x(r)\}$ is a *characteristic curve* or *characteristic*. The partial solutions $x(r)$ and $u(r)$ of (1.1.2) are sometimes referred to as the *characteristic base curve* and the *characteristic surface curve* of (1.1.1).

Thus, by definition, the characteristic curve through a point of an integral surface is tangent to the surface at that point. In fact, the following stronger theorem (see N.1.2) shows that it remains imbedded in the integral surface.

THEOREM 1.1.1. Let the vector-valued functions F, G, and w be locally Lipschitz continuous in all arguments; then w is a solution of (1.1.1) if and only if $w(x(r)) - u(r) \equiv 0$ along the characteristic $\{u(r), x(r)\}$.

Proof. Let w be a solution of (1.1.1) in a neighborhood of $x_0 \in E^n$. Then the initial value problem

$$x' = G(w(x), x), \qquad x(r_0) = x_0$$

has a unique local solution $x(r)$. Differentiation and the property that w is an integral surface show that

$$\frac{d}{dr} w(x(r)) = w_x(x)x'(r) = w_x(x)G(w, x) = F(w, x)$$

so that $\{w(r), x(r)\}$ is a characteristic. Under the continuity hypotheses the characteristic through $(w(x_0), x_0)$ is unique, so that $w(x(r)) \equiv u(r)$ in a neighborhood of r_0. Conversely, suppose that $w(x(r)) - u(r) \equiv 0$. Then $0 \equiv dw/dr - du/dr = w_x(x)G(w, x) - F(u, x) = w_x(x)G(w, x) - F(w, x)$ from which it follows that $w(x)$ is a solution of (1.1.1) over $x(r)$. ∎

In general there will exist uncountably many integral surfaces for (1.1.1) through any given point. However, if the surface is required to pass through a given curve C which is not characteristic, then the integral surface for (1.1.1) exists and is unique in a neighborhood of C. The following theorem makes this statement precise, and its proof will indicate how the characteristic equations can be used to generate an integral surface through a given manifold.

THEOREM 1.1.2. Let the initial manifold C in (u, x)-space be given parametrically by $u = f(s)$, $x = g(s)$, where $s = (s_1, \ldots, s_{n-1})^T$ is an $n - 1$ dimensional parameter, and assume that F, G, f, and g are continuously differentiable. If the Jacobian

$$\Delta \equiv \partial(x_1, \ldots, x_n)/\partial(r, s_1, \ldots, s_{n-1})$$

is nonzero on C ($\Delta \neq 0$ for all s), then the initial value problem (or Cauchy problem)

$$u_x G(u, x) = F(u, x), \qquad u(g(s)) = f(s)$$

has a unique solution in a neighborhood of C.

Proof. Under the continuity hypotheses the initial value problem

$$u'(r) = F(u, x), \qquad u(0) = f(s)$$
$$x'(r) = G(u, x), \qquad x(0) = g(s)$$

has a unique solution $(u(r, s), x(r, s))$ which is differentiable with respect to s. By hypothesis the Jacobian

$$\Delta \equiv \partial(x_1, \ldots, x_n)/\partial(r, s_1, \ldots, s_{n-1})$$

does not vanish on C. Hence the implicit function theorem assures the existence of the inverse functions $s = s(x)$, $r = r(x)$ of $x = x(r, s)$ in a neighborhood of C. Substitution of s and r into $u(r, s)$ and differentiation show that

$$\frac{d}{dr} u(r(x), s(x)) = u_x(x) \frac{dx}{dr} = u_x(x) G(u, x) = F(u, x)$$

Consequently, $u(x) \equiv u(r(x), s(x))$ is an integral surface of (1.1.1), while its construction assures that $u(g(s)) = f(s)$. Moreover, since the characteristic through a given point and the inverse functions are unique, there can exist only one integral surface through the given manifold. ∎

The condition $\Delta \not\equiv 0$ rules out the case where C itself is a characteristic. For if C were characteristic, we can choose the free parameter s_1 to vary such that $\partial x / \partial r = \partial x / \partial s_1$, which in turn forces $\Delta \equiv 0$. Geometrically, the existence of infinitely many integral surfaces through a characteristic curve is apparent because each noncharacteristic manifold crossing C allows the construction of an integral surface containing C. In subsequent applications the stronger condition $\Delta \neq 0$ will always be satisfied.

Use of the Jacobian necessarily restricts the applicability of Theorem 1.1.2 to finite dimensional systems of equations. However, it is well known that the implicit function theorem is equally valid in general function spaces, and it is not difficult to rephrase the theorem for a more abstract setting (see Chapter 5). Moreover, the proof of Theorem 1.1.2 is constructive and, under proper restrictions on the growth of the functions F and G, can be used to give a quantitative estimate of the region of existence for the integral surface. In order to relate these results more readily to the discussion of boundary value problems in subsequent chapters, we shall write the first-order equation with the same principal part (1.1.1) in the form

$$u_t + u_x(t, x) G(t, u, x) = F(t, u, x) \tag{1.1.3}$$

where t is a scalar parameter. We note that Eq. (1.1.1) can always be transformed into (1.1.3) simply by adding the new independent variable t with the trivial characteristic equation $dt/dr = 1$. The full set of characteristic equations for (1.1.3) is given by

$$du/dr = F(t, u, x), \qquad dx/dr = G(t, u, x), \qquad dt/dr = 1$$

We shall identify t with r and discard the last equation. Thus, when we speak of a characteristic for (1.1.3), we shall always think of $\{u(t), x(t)\}$, while $x(t)$ is the corresponding base characteristic. In addition, it will be assumed that the initial manifold C is of the (parametric) form $u(0) = f(s)$, $x = s$, $t = 0$ so that the Jacobian on C satisfies

$$\Delta = \frac{\partial(t, x_1, \ldots, x_n)}{\partial(r, s_1, \ldots, s_n)} = \frac{\partial(x_1, \ldots, x_n)}{\partial(s_1, \ldots, s_n)} = 1$$

Hence Theorem 1.1.2 always will yield a local integral surface for (1.1.3). Let us next give a quantitative version of this result for systems of characteristic equations which are close to linear (N.1.3). We shall present this theorem in detail because of its intimate connection with the existence of solutions for two-point boundary value problems as discussed in the next chapter.

THEOREM 1.1.3. Let F, G, and f be continuously differentiable in all variables and assume that there exist constants a, b, c, d, and i such that for $t \in [0, T]$ we have $\| F_u \| \leq a$, $\| F_x \| \leq b$, $\| G_u \| \leq c$, $\| G_x \| \leq d$, and $\| f_x \| \leq i$ uniformly in u and x. Then the Cauchy problem

$$u_t(t, x) + u_x(t, x)G(t, u, x) = F(t, u, x), \qquad u(0, x) = f(x) \quad (1.1.4)$$

has a unique integral surface $u(t, x)$ defined for all x and for $t \in [0, \hat{t})$, where $\hat{t}$ is given by

$$\hat{t} = \frac{1}{k + d} \ln\left(1 + \frac{k + d}{c \max\{1, i\}}\right) \tag{1.1.5}$$

with $k = \max\{a + b, c + d\}$.

Proof. As outlined in the proof of the Theorem 1.1.2, the integral surface can be constructed from the characteristic curves through the initial manifold $u(0, s) = f(s)$. Because of the boundedness of the derivatives of F and G, the characteristics exist over $(0, \infty)$. The next problem is to find the inverse function $s = s(t, x)$ of the base characteristic $x = x(t, s)$. (We recall from Theorem 1.1.2 that if $s = s(t, x)$ exists then $u(t, s(t, x))$ is a solution of (1.1.4).) It is well known that this function exists in a neighborhood of the characteristic provided the Jacobian

$$\Delta = \partial(x_1, \ldots, x_n)/\partial(s_1, \ldots, s_n)$$

does not vanish on $(u(t, s), x(t, s))$. This will be the case if the matrix $x_s(t, s)$ is nonsingular. Because F, G, and f are continuously differentiable, it follows that u_s and x_s satisfy the differential equations

$$u_s' = F_u u_s + F_x x_s, \qquad u_s(0) = f_s(s)$$
$$x_s' = G_u u_s + G_x x_s, \qquad x_s(0) = I \tag{1.1.6}$$

where I is the identity matrix.

The solution of this linear system may be written as

$$u_s = \phi(t, 0)f_s + \int_0^t \phi(t, r)F_x x_s(r)\, dr$$

$$x_s = \psi(t, 0)I + \int_0^t \psi(t, r)G_u u_s(r)\, dr$$

where, for given $\{u(t), x(t)\}$, ϕ and ψ are (invertible) fundamental matrices of the systems

$$\phi' = F_u(t, u(t), x(t))\phi, \qquad \phi(0) = I$$
$$\psi' = G_x(t, u(t), x(t))\psi, \qquad \psi(0) = I$$

The above representation shows that x_s is invertible if the operator

$$\psi(t, 0)\left[I + \int_0^t \psi(0, r)G_u u_s(r)\, dr \right]$$

is invertible. Banach's lemma assures that this is the case whenever

$$\left\| \int_0^t \psi(0, r)G_u u_s\, dr \right\| < 1 \tag{1.1.7}$$

Hence an estimate for u_s is needed. Applying Gronwall's inequality to (1.1.6) and using the hypotheses, we obtain the inequality

$$\max\{\| u_s(t) \|,\ \| x_s(t) \|\} \le \max\{1, i\}e^{kt}$$

with $k = \max\{a + b, c + d\}$. Since also $\| \psi(0, r) \| \le e^{dr}$ we can estimate (1.1.7);

$$\left\| \int_0^t \psi(0, r)G_u u_s\, dr \right\| \le c \max\{1, i\} \int_0^t e^{(k+d)r}\, dr$$

$$= c \max\{1, i\} \frac{1}{k + d} (e^{(k+d)t} - 1)$$

Hence (1.1.7) is satisfied provided t is chosen smaller than $\hat{t}$, where

$$\hat{t} = \frac{1}{k + d} \ln\left(1 + \frac{d + k}{c \max\{1, i\}}\right)$$

Thus, for $t < \hat{t}$ we can find the inverse function $s = s(t, x)$ of the base characteristic $x = x(t, s)$, which assures the existence of the integral surface $u(t, x)$ in a neighborhood of the characteristic. Moreover, because x_s is nonsingular, $x = x(t_0, s)$ is a homeomorphism on E^n for each $t_0 < \hat{t}$, so that a base characteristic passes through each point of E^n as s ranges over E^n. As a result, the integral surface $u(t, x)$ just constructed exists for all x and for $t < \hat{t}$. ∎

The use of norm estimates and Banach's lemma will disguise any structure the system (1.1.6) may possess. The crude estimate (1.1.5) does, however, account for the uncoupling which occurs when $c \to 0$. Not accounted for is the uncoupling which occurs when both i and b approach zero. The bound (1.1.5) can be sharpened to reflect this behavior by iterating on the bound for $\| u_s \|$. It may be verified that substitution of the bound $\| x_s \| = \max\{1, i\}e^{kt}$ into the expression for $\| u_s \|$ and use of Gronwall's inequality yield the new bound

$$\| u_s \| \le \left[i + \max\{i, 1\}b\left(\frac{e^{kt} - 1}{k}\right) \right]e^{at}$$

which on substitution into (1.1.7), requires that t be chosen so small that

$$\frac{ci}{a + d}(e^{(a+d)t} - 1) + \max\{i, 1\}\frac{cb}{k}\left[\frac{e^{(k+d+a)t} - 1}{k + d + a} - \frac{e^{(a+d)t} - 1}{a + d}\right] < 1$$

In order to solve this inequality for t we shall increase the first and neglect the last term to obtain

$$\left[\frac{ci}{a + d} + \max\{i, 1\}\frac{cb}{k(k + d + a)}\right](e^{(k+d+a)t} - 1) < 1$$

From this inequality one can compute a maximum permissible $\hat{t}$ as

$$\hat{t} = \frac{1}{k + d + a}\ln\left(1 + \left\{\frac{ci}{a + d} + \max\{i, 1\}\frac{cb}{k(k + d + a)}\right\}^{-1}\right) \quad (1.1.8)$$

Note that $\hat{t} \to \infty$ as $c \to 0$ or $i \to 0$ and $b \to 0$ (see N.1.4).

The bounds in Theorem 1.1.3 were required to hold uniformly in u and x. Using the next lemma we can relax these conditions somewhat (see N.1.5).

LEMMA 1.1.1. Let $D_1 = \{x: \| x \| \le \alpha\}$, $D_2 = \{x: \| x \| \le \beta\}$

$$0 < \delta = \beta - \alpha$$

Then the function h defined on E^n as

$$h(x) = \begin{cases} x, & x \in D_1 \\ (\beta - \delta e^{-(\|x\|-\alpha)/\delta})(x/\|\,x\,\|), & x \notin D_1 \end{cases}$$

is continuously differentiable on E^n and satisfies $\|\,h_x(x)\,\| \leq 1$.

Proof. For ease of notation set $r = \|\,x\,\|$ and $s(r) = (1/r)(\beta - \delta e^{-(r-\alpha)/\delta})$; then $h(x) = s(\|\,x\,\|)x$. By inspection it is seen that h is continuous on E^n and assumes values in D_2. It also is continuously differentiable on E^n for the norm is differentiable and satisfies $\|\,x\,\|_x z = \langle x, z \rangle/\|\,x\,\|$ for arbitrary $z \in E^n$. Consequently, the chain rule may be applied to h. Straightforward computation shows that

$$h_x(x)z = (s'(r)(\langle x, z \rangle/r)x + s(r)z) \tag{1.1.9}$$

where $s'(r) = -(1/r)s(r) + (1/r)e^{-(r-\alpha)/\delta}$. By inspection $s'(\alpha) = 0$ and $s(\alpha) = 1$ so that $h_x(x)z = z$ when $\|\,x\,\| = \alpha$. This establishes continuity of h_x. Finally, we note from (1.1.9) that for $\|\,x\,\| \geq \alpha$

$$\|\,h_x(x)z\,\|^2 = s^2(r)\left(1 - \frac{\langle x, z \rangle^2}{r^2}\right) + \frac{\langle x, z \rangle^2}{r^2}\, e^{-2(r-\alpha)/\delta}$$

Since $-e^{-\varepsilon} \leq \varepsilon - 1$ for $\varepsilon > 0$ we see that

$$s(r) \leq (1/r)[\beta + (\{r - \alpha\}/\delta - 1)\delta] = 1$$

for $r \geq \alpha$ which in turn implies that $\|\,h_x(x)\,\| \leq 1$. ∎

This lemma can be used in conjunction with Theorem 1.1.3 to give the following result.

THEOREM 1.1.4. Let F, G, and f satisfy the conditions of Theorem 1.1.3 on the set $R \times E^m \times D_2$. Then there exists an integral surface $u(t, x)$ for (1.1.4) over the strip $[0, \hat{t}) \times D_1$, where $\hat{t}$ is given by (1.1.5) (or (1.1.8)).

Proof. Define $F_1(t, u, x) = F(t, u, h(x))$, $G_1(t, u, x) = G(t, u, h(x))$, and $f_1(x) = f(h(x))$. Then Theorem 1.1.3 assures the existence of an integral surface $u_1(t, x)$ over $[0, t) \times E^n$ for the Cauchy problem.

$$u_t(t, x) + u_x(t, x)G_1(t, u, x) = F_1(t, u, x), \qquad u(0, x) = f_1(x)$$

Since $h(x) = x$ on D_1, it follows that $u_1(t, x)$ is a solution of (1.1.4) on $[0, \hat{t}) \times D_1$. ∎

In general, the region of existence for $u(t, x)$ may or may not extend far beyond $\hat{t}$. For example, the scalar-valued Cauchy problem

$$u_t(t, x) + u_x(t, x)u = x, \qquad u(0, x) = 0$$

has the global solution $u(t, x) = x \tanh t$, while the Cauchy problem

$$u_t(t, x) - u_x(t, x)u = x, \qquad u(0, x) = 0$$

has the local solution $u(t, x) = x \tan t$. For both equations, Theorem 1.1.4 assures the existence of u for all x and for $t < \hat{t} = \ln 2$. Incidentally, the same result is obtained from the iterated bound provided we take the proper limits before estimating $\hat{t}$. (This becomes necessary because $a = d = 0$.)

1.2. LINEAR CHARACTERISTIC EQUATIONS

If the characteristic equations of a system of partial differential equations with the same principal part are linear inhomogeneous equations, then an important simplification of the foregoing theory occurs. The partial differential equations can now be reduced to ordinary differential equations of the Riccati type. Since this equivalence is fundamental in the discussion of boundary value problems for linear equations, we shall develop the simplified theory in detail.

Let us suppose that the characteristic equations are given as

$$\begin{aligned} u'(t) &= A(t)u + B(t)x + F(t) \\ x'(t) &= C(t)u + D(t)x + G(t) \end{aligned} \qquad (1.2.1)$$

where $u = (u_1, \ldots, u_m)^T$, $x = (x_1, \ldots, x_n)^T$, and where A, B, C, D are matrices of consistent dimensions. All given functions are assumed to be continuous. The extension of the theory to differential equations with only measurable data functions and Caratheodory-type solutions is discussed in a coordinate-free setting in Chapter 5. According to Theorem 1.1.1, the characteristic $\{u(t), x(t)\}$ can be imbedded in an integral surface $u(t, x)$ of the partial differential equation.

$$u_t(t, x) + u_x(t, x)[C(t)u + D(t)x + G(t)] = A(t)u + B(t) + F(t) \quad (1.2.2)$$

Let us require that the integral surface pass through a linear initial manifold of the form

$$u(0, x) = fx + a$$

where f is an $m \times n$ matrix and $a = (a_1, \ldots, a_m)^T$ is a given vector. According to Theorem 1.1.2, this integral surface can be generated with the characteristics. As outlined in the proof of Theorem 1.1.3, we integrate the initial value problem

$$u'(t) = A(t)u + B(t)x + F(t), \qquad u(0) = fs + a$$
$$x'(t) = C(t)u + D(t)x + G(t), \qquad x(0) = s$$

Its solution is given by

$$\begin{pmatrix} u(t, s) \\ x(t, s) \end{pmatrix} = \phi(t, 0)\begin{pmatrix} fs + a \\ s \end{pmatrix} + \int_0^t \phi(t, r)\begin{pmatrix} F(r) \\ G(r) \end{pmatrix} dr \qquad (1.2.3)$$

where $\phi(t, r)$ is the fundamental solution of the $(m + n) \times (m + n)$ system

$$C' = \begin{pmatrix} A(t) & B(t) \\ C(t) & D(t) \end{pmatrix} C, \qquad C(0) = I$$

In component form, Eqs. (1.2.3) may be written

$$u(t, s) = \phi_{11}(t, 0)(fs + a) + \phi_{12}(t, 0)s + \hat{F}(t)$$
$$x(t, s) = \phi_{21}(t, 0)(fs + a) + \phi_{22}(t, 0)s + \hat{G}(t)$$

where $\hat{F}$ and $\hat{G}$ denote the first m and remaining n equations of the last term of (1.2.3). As outlined above, we have to find the inverse function $s = s(t, x)$, which in this case is readily computed to be

$$s = [\phi_{21}(t, 0)f + \phi_{22}(t, 0)]^{-1}[x - \phi_{21}(t, 0)a - \hat{G}(t)]$$

Note that the initial condition $C(0) = I$ requires that $\phi_{21}(0, 0) = 0$ and $\phi_{22}(0, 0) = I$, so that s is defined in a neighborhood of $t = 0$. Of course, the same conclusion is also assured by Theorem 1.1.1. Substitution of s into $u(t, s)$ yields

$$u(t, x) = [\phi_{11}(t, 0)f + \phi_{12}(t, 0)]\{[\phi_{21}(t, 0)f + \phi_{22}(t, 0)]^{-1}x$$
$$- [\phi_{21}(t, 0)f + \phi_{22}(t, 0)]^{-1}[\phi_{21}(t, 0)a + \hat{G}(t)]\} + \phi_{11}(t, 0)a + \hat{F}(t)$$

Thus, we see that $u = u(t, x)$ has the representation

$$u(t, x) = U(t)x + v(t) \qquad (1.2.4)$$

For ease of reference we shall call the expression (1.2.4) an *affine transformation* although U will generally not be invertible. If this representation is substituted into (1.2.2), the following equation results:

$$U'(t)x + v'(t) + U[C(t)Ux + C(t)v + D(t)x + G(t)]$$
$$= A(t)Ux + A(t)v + B(t)x + F(t)$$

This expression has to hold for all x so that it can be separated into the system

$$[U'(t) + UC(t)U + UD(t) - A(t)U - B(t)]x = 0$$
$$v'(t) + U(t)C(t)v + U(t)G(t) - A(t)v - F(t) = 0$$

The first of these equations is known as the matrix Riccati equation. The initial condition $u(0, x) = fx + a$ requires that $U(0)x = fx$ and $v(0) = a$ for all x. Conversely, it follows by differentiation and use of the defining equations for U and v that the function $u(t, x) = U(t)x + v(t)$ is the necessarily unique solution of (1.2.2) because $u(0, x) = fx + a$. We shall summarize this discussion as

THEOREM 1.2.1. The integral surface $u(t, x)$ of the Cauchy problem
$$u_t(t, x) + u_x(t, x)[C(t)u + D(t)x + G(t)] = A(t)u + B(t)x + F(t)$$

$$u(0, x) = fx + a$$

has the representation $u(t, x) = U(t)x + v(t)$, where the $m \times n$ matrix U satisfies the ordinary differential Riccati equation

$$U'(t) = B(t) + A(t)U - UD(t) - UC(t)U, \qquad U(0) = f \qquad (1.2.5)$$

while $v(t)$ is the solution of the linear equation

$$v'(t) = -[U(t)C(t) - A(t)]v - U(t)G(t) + F(t), \qquad v(0) = a \qquad (1.2.6)$$

Note that the matrices A and D in (1.2.5) must be square matrices of order $m \times m$ and $n \times n$, respectively, since otherwise the dimensions are inconsistent. For the same reason B and C must be of order $m \times n$ and $n \times m$, respectively.

We also remark that the discussion leading to Theorem 1.2.1 presents a geometric motivation for the well-known reduction of the second-order differential equation

$$u'' + p(t)u' + g(t)u = 0$$

to a first-order scalar Riccati equation

$$v' + v^2 + p(t)v + g(t) = 0$$

through the so-called Riccati substitution

$$vu' = u$$

In fact, writing the second-order equation as a first-order system, we see that the Riccati substitution is precisely the affine transformation (1.2.4) relating u and u'. Consequently, (1.2.4) may be considered a generalized Riccati transformation for inhomogeneous equations.

As an illustration, let us consider the two examples discussed earlier. Identification of coefficients shows that the Cauchy problem

$$u_t(t, x) + u_x(t, x)u = x, \qquad u(0, x) = 0$$

leads to the Riccati equation

$$U'(t) = 1 - U^2, \qquad U(0) = 0 \tag{1.2.7}$$

which has the solution $U(t) = \tanh t$, while the Cauchy problem

$$u_t(t, x) - u_x(t, x)u = x, \qquad u(0, x) = 0$$

corresponds to the Riccati equation

$$U'(t) = 1 + U^2, \qquad U(0) = 0 \tag{1.2.8}$$

Its solution is $U(t) = \tan t$. The linear equation is $v' = -U(t)v$ in the first and $v' = U(t)v$ in the second case. Since $v(t) = v(0) \equiv 0$, the representation $u(t, x) = U(t)x + v(t)$ reduces to the solutions given previously.

1.3. THE RICCATI EQUATION

We shall see later that linear boundary value problems require the solution of a matrix Riccati differential equation and that the existence of solutions for the Riccati equation is readily translated into existence theorems for the boundary value problems. We shall therefore discuss Eq. (1.2.5) in more detail. To make this section self-contained, let us restate this equation. We shall deal with the following $m \times n$ matrix differential equation

$$U' = B(t) + A(t)U - UD(t) - UC(t)U, \qquad U(0) = f \tag{1.3.1}$$

Because of its natural connection with the first-order linear system (1.2.1), we have retained the negative sign for the last two terms of the equation. This is at variance with some of the published work on matrix Riccati equations so that a comparison with other results may require adjusting the algebraic signs.

There exists a considerable body of literature on the scalar and matrix Riccati equation, and we refer to the monograph of Reid (1972) for a comprehensive study of its properties and its role in the theory and application of ordinary differential equations. The material here summarizes those features of the general theory which bear on the development of invariant imbedding.

For linear systems like (1.2.1) with continuous coefficients, the corresponding Riccati equation will always have a local solution for sufficiently small t. In fact, without further specifying the structure of (1.3.1), we can rephrase Theorem 1.1.3 to obtain a quantitative existence theorem.

THEOREM 1.3.1. Let a, b, c, d, and i be constants such that $\| A(t) \| \leq a$, $\| B(t) \| \leq b$, $\| C(t) \| \leq c$, $\| D(t) \| \leq d$, and $\| f \| \leq i$ on the interval $[0, T]$. Then the Riccati equation (1.3.1) has a solution for all t such that $t \in [0, T]$ and

$$t < \hat{\imath} = \frac{1}{k + d} \ln\left(1 + \frac{k + d}{c \max\{1, i\}}\right)$$

where $k = \max\{a + b, \ c + d\}$.

Proof. It follows from (1.2.3) that

$$u(t, x) = u_s(t)x_s(t)^{-1}x + [\hat{F}(t) - x_s(t)^{-1}\hat{G}(t)]$$

so that $U(t) = u_s(t)x_s(t)^{-1}$. The bound of the theorem was calculated such that x_s^{-1} exists on compact subsets of $[0, \hat{\imath})$. ∎

Given the local existence of the solution $U(t)$ for (1.3.1), the problem now centers around the question of how far this solution can be continued. For Hermitian Riccati equations, a reasonably complete answer can be given. The Hermitian version of (1.3.1) will be written as

$$U' = B(t) + A(t)U + UA^*(t) - UC(t)U, \qquad U(0) = f \qquad (1.3.2)$$

where B, C, and f are Hermitian matrices, and where $*$ denotes the Hermitian adjoint. As pointed out already, (1.3.2) has a unique local solution U near 0. It is also readily verified that U^* satisfies (1.3.2) as well. Consequently, the solution of the Hermitian Riccati equation is itself Hermitian.

Equation (1.3.2) has a very special form, but it arises naturally in control theory when optimizing the quadratic performance functional of a linear regulator (see Examples 2.10.1 and 5.3.1), and when solving numerically certain elliptic boundary value problems (Example 2.8.2). Many technical applications lead to such problems and as a result the Hermitian Riccati equation has been studied quite extensively (N.3.1). We shall offer here an additional tool for the study of Riccati equations which will allow us to rederive and extend some of the well-known results reported in the literature, and which holds some promise for the study of non-Hermitian equations.

Our existence and comparison theorems for (1.3.2) will be based on an elementary lemma. For its discussion, it will be convenient to introduce some additional notation. Subsequently M shall denote the space of Hermitian operators defined on E^n, while M_0 designates the subspace of nonnegative Hermitian operators. While the setting of these results is the finite-dimensional space E^n, it should be noted that the tools employed apply equally well in an arbitrary inner product space. We shall return to the infinite dimensional aspects of this theory in Chapters 5 and 6.

LEMMA 1.3.1. Let $F: [0, T] \times M \to M$ be continuous in t and locally Lipschitz continuous in X. Suppose further that for arbitrary $X \in M$ and $t \in [0, T]$ we have $\langle F(t, X)u_0, u_0 \rangle \geq 0$ whenever $u_0 \neq 0$ belongs to the null space $\nu(X)$ of X. If the solution $X(t)$ of

$$X' = F(t, X), \qquad X(0) = X_0$$

exists over $[0, T]$, then $X_0 \in M_0$ implies that $X(t) \in M_0$ for all $t \in [0, T]$.

Proof. Consider the solution $X_\varepsilon(t)$ of

$$X_\varepsilon' = F(t, X_\varepsilon), \qquad X_\varepsilon(0) = X_0 + \varepsilon I$$

for an arbitrary $\varepsilon > 0$. Since $\langle X_\varepsilon(0)u, u \rangle > 0$ for all $u \in E^n$, it follows that $\langle X_\varepsilon(t)u, u \rangle > 0$ over some nonzero interval. Let t_0 be the smallest value of t such that $\langle X_\varepsilon(t_0)u_0, u_0 \rangle = 0$ for some nonzero u_0. Because of the hypothesis on F, we obtain $\langle X_\varepsilon(t_0)u_0 u_0 \rangle' \geq 0$. This condition is inconsistent with $\langle X_\varepsilon(t)u_0, u_0 \rangle > 0$ on $[0, t_0)$. Hence $X_\varepsilon(t)$ is positive definite on $[0, T]$. Continuity of the solution X_ε in ε allows the conclusion that

$$\lim_{\varepsilon \to 0} \langle X_\varepsilon(t)u, u \rangle = \langle X(t)u, u \rangle \geq 0. \quad \blacksquare$$

COROLLARY 1.3.1. If B and f are nonnegative, then the solution $U(t)$ of the Hermitian Riccati equation (1.3.2) is nonnegative for $t \geq 0$.

Proof. Denoting the right-hand side of (1.3.2) by $F(t, U)$, we see that $\langle F(t, U)u_0, u_0 \rangle = \langle B(t)u_0, u_0 \rangle \geq 0$ for $u_0 \in v(U)$. Hence, Lemma 1.3.1 applies. ∎

COROLLARY 1.3.2. *If there exists an α such that $\langle (\alpha I - f)u, u \rangle \geq 0$ and*

$$\alpha^2 \langle Cu, u \rangle - \langle Bu, u \rangle - \alpha \langle (A + A^*)u, u \rangle \geq 0 \tag{1.3.3}$$

then

$$\langle [\alpha I - U(t)]u, u \rangle \geq 0$$

Proof. Let $X(t) = \alpha I - U(t)$; then it follows from (1.3.2) that

$$X' = -B(t) - \alpha[A(t) + A^*(t)] + \alpha^2 C(t)$$
$$+ [A(t) - \alpha C(t)]X + X[A^*(t) - \alpha C(t)] + XC(t)X$$

By hypothesis $\langle X(t)u_0u_0 \rangle' \geq 0$ whenever $u_0 \in v(X(t))$ and the conclusion follows from Lemma 1.3.1. ∎

We note that a positive α satisfying (1.3.3) can always be found whenever C is positive definite. In fact, Lemma 1.3.1 and Corollary 1.3.2 together furnish an elementary proof of the well-known result that the Hermitian Riccati equation maps the cone of positive definite operators into itself (see N.3.2). But even if C is negative definite the Riccati equation may have bounded positive solutions provided the linear terms sufficiently retard the growth of the solution. Consider, for example, the one-dimensional equation

$$x' = 1 - \gamma x + x^2, \qquad x(0) = 0$$

We have seen above that for $\gamma = 0$, the solution is $x(t) = \tan t$. It may be noted that for $\gamma \neq 0$ the above equation has a bounded solution whenever $x' = 0$ has a positive solution x. This requires that $\gamma \geq 2$, and the same conclusion is obtained if we evaluate the inequality (1.3.3).

Corollary 1.3.2 is actually a comparison theorem, and an analogous proof can be used to compare the solutions of different Riccati equations.

COROLLARY 1.3.3. *Assume that the two Riccati equations*

$$U_i' = B_i(t) + A(t)U_i + U_iA^*(t) - U_iC_i(t)U_i, \qquad U_i(0) = f_i, \quad i = 1, 2$$

have bounded solutions over $[0, T]$. Suppose further that $B_2(t) - B_1(t) \in M_0$, $C_1(t) - C_2(t) \in M_0$, and $f_2 - f_1 \in M_0$, then $U_2(t) - U_1(t) \in M_0$.

Proof.

$$\langle [U_2(t) - U_1(t)]u_0, u_0 \rangle' = \langle [B_2(t) - B_1(t)]u_0, u_0 \rangle$$
$$+ \langle [C_1(t) - C_2(t)]U_1(t)u_0, U_1(t)u_0 \rangle \geq 0$$

whenever $u_0 \neq 0$ belongs to the null space of $U_2(t) - U_1(t)$. Hence Lemma 1.3.1 applies. ∎

Precise bounds on the solution of (1.3.2) are obtainable if we can compare with known solutions of related Riccati equations. Jacobson (1970) exhibits a case where the matrix C can be factored into $C(t) = D(t)R(t)D^*(t)$ and where an initial value f_2 and an inhomogeneous term $B_2(t)$ satisfying the conditions of Corollary 1.3.3 can be found such that

$$D^*(t)U_2(t) \equiv 0 \qquad \text{on} \quad [0, T]$$

In this case a bound on U_1 can be calculated by solving the linear equation

$$U_2' = B_2(t) + A(t)U_2 + U_2 A^*(t)$$

which always has a global solution on $[0, \infty)$.

If $A(t) \equiv 0$ and upper and lower bounds on B and C are readily obtainable, we can find explicit upper and lower bounds. This is due to the observation that the diagonal Riccati equation

$$U' = bI - cU^2, \qquad U(0) = f \tag{1.3.4}$$

has the closed form solution

$$U(t) = [\alpha(e^{\alpha t} + e^{-\alpha t})I + cf(e^{\alpha t} - e^{-\alpha t})]^{-1}$$
$$\times [\alpha f(e^{\alpha t} + e^{-\alpha t}) + b(e^{\alpha t} - e^{-\alpha t})I] \tag{1.3.5}$$

where $\alpha = (bc)^{1/2}$. Suppose, for example, that

$$0 \leq m_1 \leq \langle B(t)u, u \rangle \leq M_1 \qquad \text{and} \qquad 0 \leq m_2 \leq \langle C(t)u, u \rangle \leq M_2$$

for all u satisfying $\| u \| = 1$.

It now follows from Corollary 1.3.3 that the solution U of (1.3.2) (with $A(t) \equiv 0$) remains bounded on $[0, T]$ provided

$$\langle fu, u \rangle \geq -K \| u \|^2$$

for all $u \in E^n$ where

$$K = \max \left\{ \frac{\sqrt{M_1}}{\sqrt{m_2} \tanh(M_1 m_2)^{1/2}T}, \frac{\sqrt{m_1}}{\sqrt{M_2} \tanh(M_2 m_1)^{1/2}T} \right\}$$

Indeed, it is a straightforward calculation to establish that with this choice of K the upper and lower bounding solution for (1.3.2) obtained from (1.3.5) will remain finite.

As a second example, suppose that m_1 is positive and M_2 is negative. For simplicity suppose also that $f = 0$. Then the upper bound on $U(t)$ (again with $A(t) \equiv 0$) will remain finite provided $| \tanh(M_1 m_2)^{1/2} t | = \tan(| M_1 m_2 |) t$ remains bounded on $[0, T]$.

We shall leave the topic of Hermitian Riccati equations at this point and refer the reader to the literature cited at the end of this chapter for a more extensive discussion of Hermitian matrix Riccati differential equations. Here, we shall examine to what extent the above approach is applicable to the general Riccati equation (1.3.1). We shall assume now that U is an $m \times n$ matrix and that the coefficient matrices are complex, continuous and of consistent dimensions.

Rather than bounding the numerical range $\langle U(t)x, x \rangle$ as was possible for Hermitian U, we shall work with the norm of U defined by

$$\| U \| = \sup_{\|x\|=\|y\|=1} |\langle Ux, y \rangle| = \max_{\|x\|=\|y\|=1} |\langle Ux, y \rangle|$$

Note that x and y need not be of identical dimension. The connection to the above results is apparent because for Hermitian U we can choose $x = y$. The following generalization of Corollary 1.3.2 can be obtained (see also N.3.3).

THEOREM 1.3.2. Suppose that $\| f \| < 1$ and that the Hermitian form

$$\Omega = \langle (B^*(t) - C(t))y, x \rangle + \langle (B(t) - C^*(t))x, y \rangle + \langle (A(t) + A^*(t))y, y \rangle$$
$$- \langle (D(t) + D^*(t))x, x \rangle$$

is negative for all unit vectors x and y and all $t \in [0, T]$. Then the solution $U(t)$ of (1.3.1) exists and satisfies $\| U(t) \| < 1$ on $[0, T]$.

Proof. Suppose that $\| U(t) \| < 1$ on $[0, \hat{t})$ and that $\| U(\hat{t}) \| = 1$. Then we can find unit vectors $\hat{x}$ and $\hat{y}$ such that $\langle U(t)\hat{x}, \hat{y} \rangle = 1$. If $U(t)\hat{x}$ is expressed as $U(\hat{t})\hat{x} = \alpha\hat{y} + r$ where $\alpha\hat{y}$ is the projection of $U(\hat{t})\hat{x}$ onto $\hat{y}$ and r its orthogonal complement, then it follows from $\langle U(\hat{t})\hat{x}, \hat{y} \rangle = 1$ that $\alpha = 1$. Moreover, since $\| U(\hat{t})\hat{x} \| \leq \| U(\hat{t}) \| = 1$, the vector r must vanish. Hence at this point $U(\hat{t})\hat{x} = \hat{y}$. Next we observe from the hypotheses that the complex function $h(t) = \langle U(t)\hat{x}, \hat{y} \rangle$ describes a curve lying inside the unit disk for $t \in [0, \hat{t})$ and touching the boundary at $z = 1$ when $t = \hat{t}$. This implies that there exists a neighborhood bounded above by $\hat{t}$ such that

Re $h'(t) \geq 0$ for t in this neighborhood. It is straighforward to verify from $U(\hat{t})\hat{x} = \hat{y}$ and Re $h'(t) = \frac{1}{2}[\langle U(t)\hat{x}, \hat{y} \rangle + \langle \hat{y}, U(t)\hat{x} \rangle]$ that Re $h'(t) = \frac{1}{2}\Omega$. By hypothesis $\Omega < 0$, which contradicts the requirement that Re $h'(t) \geq 0$. ∎

Let us look at a consequence of the condition $\Omega < 0$. Suppose that we are dealing only with real matrices and vectors. Then the inequality for Ω can be reduced to

$$\Omega_1 = \langle (B - C^*)x, y \rangle + \langle Ay, y \rangle - \langle Dx, x \rangle < 0$$

If $D(t)$ and $-A(t)$ are bounded from below by nonzero constants a and d such that $\langle -A(t)y, y \rangle \geq a\langle y, y \rangle$ and $\langle D(t)x, x \rangle \geq d\langle x, x \rangle$ for arbitrary x and y, then Ω_1 can be estimated by

$$\Omega_1 \leq \| B(t) - C^*(t) \| - a - d$$

As a consequence, the Hermitian form will always be negative provided A and D are sufficiently strong to retard the growth of an initially small solution for the Riccati equation (1.3.1).

We note also that Corollary 1.3.2 for the Hermitian Riccati equation results if the proof of Theorem 1.3.2 is based on the numerical range definition of the norm of U, namely $\| U \| = \sup_{\|x\|=1} \langle Ux, x \rangle$. The constant α can be included in the proof by scaling the Riccati equation (see N.3.4).

Finally, it may be noted that the transformation

$$U(t) = \phi(t, 0)Y\psi^*(t, 0)$$

where ϕ and ψ now denote the fundamental matrices of the systems

$$\phi' = A(t)\phi, \qquad \psi' = -D^*(t)\psi, \qquad \phi(0) = \psi(0) = I$$

will always reduce the general Riccati equation (1.3.1) to the expression

$$Y' = \phi(0, t)B(t)\psi^*(0, t) - Y\psi^*(t, 0)C(t)\phi(t, 0)Y, \qquad Y(0) = f$$

Hence it is possible, at least in principle, to eliminate the linear terms of the Riccati equation. This observation allows an easy proof of the positivity of solutions for certain square non Hermitian equations arising in transport theory. Such equations are characterized by the fact that B, $-C$, and f have nonnegative entries, while the off-diagonal elements of A and $-D$ are also nonnegative. Writing (1.3.1) as

$$U' = (\text{diag } A)U - U(\text{diag } D)$$
$$+ [B + (A - \text{diag } A)U - U(D - \text{diag } D) - UCU]$$

considering the bracketed term as a source term and eliminating the other linear terms, we can verify that U has only nonnegative elements. Moreover, the Hermitian form (1.3.3) again governs the existence of global solutions (see N.3.4).

1.4. NONLINEAR CHARACTERISTIC EQUATIONS

Let us return to the general nonlinear equation

$$u_t + u_x G(t, u, x) = F(t, u, x)$$
$$u(0, x) = f(x)$$

$$(1.4.1)$$

and the associated characteristic equations

$$u' = F(t, u, x), \qquad u(0) = f(s)$$
$$x' = G(t, u, x), \qquad x(0) = s$$

$$(1.4.2)$$

It was shown that if the characteristic $\{u(t), x(t)\}$ through the initial point $(f(s), s)$ exists over $[0, T]$, then the integral surface $u(t, x)$ exists in a neighborhood of this characteristic provided x_s is invertible along $x(t)$. However, u_s and x_s are solutions of the linear equations

$$u_s' = F_u(t, u(t), x(t))u_s + F_x(t, u(t), x(t))x_s, \quad u_s(0) = f_s$$
$$x_s' = G_u(t, u(t), x(t))u_s + G_x(t, u(t), x(t))x_s, \quad x_s(0) = I$$

so that x_s is invertible if and only if the Riccati equation

$$U' = F_x(t, u(t), x(t)) + F_u(t, u(t), x(t))U - UG_x(t, u(t), x(t))$$
$$- UG_u(t, u(t), x(t))U, u(0) = f_s$$

$$(1.4.3)$$

has a solution. If the characteristics exist over $[0, T]$ for all s in some connected subset, and if the Riccati equation (1.4.3) has a solution over $[0, T]$, then we can conclude from Theorem 1.1.3 that $u(t, x)$ exists over the strip containing the characteristics. Equation (1.4.3) is, of course, one of the five characteristic equations normally associated with the general nonlinear first-order partial differential equation $F(t, x, u, \partial u/\partial t, \partial u/\partial x) = 0$ (Courant and Hilbert, 1962, II, Chapter 2). It may be noted that for scalar equations the existence of a solution for (1.4.3) can sometimes be established by inspection. For example, if the initial value problem (1.4.2) has a solution for arbitrary $s \in E^n$ and if $f_s \geq 0$, $F_x \geq 0$, and $G_u \geq 0$, then the local solution of (1.4.3) will necessarily remain bounded provided the coefficients stay bounded on $[0, T]$ because $U' < 0$ as $U \to \infty$. This result is a special case of Corollary 1.3.2 and can be generalized to systems of equations. In particular, the integral surface $u(t, x)$ corresponding to the characteristic equations

$$u' = x, \qquad x' = G(t, u)$$

will exist whenever $G_u(t, u)$ is positive definite.

In theory the integral surface $u(t, x)$ of (1.4.1) is generated by the characteristics; however, in general this construction cannot be carried out in closed form unless the characteristic equations have known solutions which allow elimination of the parameter s. Occasionally, it may be possible to solve the initial value problem (1.4.1) directly with one of the many known *ad hoc* methods, such as separation of variables or coordinate transformations. In practice, these techniques reduce the partial differential equation to an ordinary differential equation much like the affine transformation (1.2.4). In fact, it is quite simple to classify those equations where a solution of the form $u(t, x) = U(t)x$ is admissible. We recall that a function f is homogeneous of degree p if $f(\lambda x) = \lambda^p f(x)$ for all x. Let us suppose now that F, G, and f are homogeneous of degree 1 in x so that

$$F(t, Ux, x) = xF(t, U, 1), \qquad G(t, Ux, x) = xG(t, U, 1), \qquad f(x) = xf(1)$$

The (necessarily unique) solution $u(t, x)$ of (1.4.1) is then given by

$$u(t, x) = U(t)x$$

where U satisfies the ordinary differential equation

$$U' = F(t, U, 1) - UG(t, U, 1), \qquad U(0) = f(1)$$

As a concrete example, we can choose

$$F(t, u, x) = [A(t)u^k + B(t)x^k]^{1/k}$$
$$G(t, u, x) = [C(t)u^m + D(t)x^m]^{1/m}$$
$$f(s) = fs$$

A number of like results are obtainable by assuming different representations for $u(t, x)$, by choosing new variables $\tau = \tau(t, x)$, $y = y(t, x)$, or possibly by introducing transformation groups for the dependent and independent variables. These techniques are described by Hansen (1964) and Ames (1967). However, there does not exist a unified theory or an adequate classification of the manageable partial differential equations, and it is to be expected, even for scalar equations, that the initial value problem (1.4.1) will only rarely admit a closed-form solution or a reduction to ordinary differential equations.

1.5. DISCRETE EVOLUTION EQUATIONS

The mechanism of relating characteristic ordinary differential equations to a system of partial differential equations with the same principal part has a direct analog when dealing with delay and difference equations. Since the resulting equations can be immediately applied to the solution of discrete boundary value problems by initial value methods (Section 2.11) we shall carry through this development in some detail. In this context we shall call

$$u(t) = F(t, \tau, u(\tau)), \qquad t \neq \tau \tag{1.5.1}$$

a *functional evolution equation*. If t and τ are restricted to a countable set, we shall speak of a *discrete evolution equation*.

Given a system of evolution equations

$$\begin{aligned} u(t) &= F(t, \tau, u(\tau), x(\tau)) \\ x(t) &= G(t, \tau, u(\tau), x(\tau)) \end{aligned} \tag{1.5.2}$$

where $u = (u_1, \ldots, u_m)^{\mathrm{T}}$, $x = (x_1, \ldots, x_n)^{\mathrm{T}}$, we can formally associate with it a functional equation of the form

$$R(t, G(t, \tau, R(\tau, x), x)) = F(t, \tau, R(\tau, x), x) \tag{1.5.3}$$

We observe from (1.5.3) that given $R(\tau, x)$ for x belonging to a known domain we can explicitly compute $R(t, z)$ for all z in the range of $G(t, \tau, R(\tau, x), x)$. It is straightforward to derive a characteristic theory for (1.5.2) and (1.5.3). For convenience we shall take $\tau = 0$. The following result is the direct analog of Theorem 1.1.1.

THEOREM 1.5.1. Let $R(t, z)$ be the solution of (1.5.3) and suppose that the evolution equations (1.5.2) have unique solutions $\{u(t), x(t)\}$ subject to a given initial point $(R(0, z_0), z_0)$. Then $\{R(t, x(t)), x(t)\}$ is a solution of (1.5.2).

Proof. We observe from the hypothesis that

$$R(t, x(t)) = R(t, G(t, 0, R(0, z_0), z_0)) = F(t, 0, u(0), x(0))$$

The uniqueness of the solutions assures that $R(t, x(t)) \equiv u(t)$. ∎

As in the case of differential equations, an "integral surface" $R(t, x)$ for (1.5.3) through a given initial manifold C can be generated with the "characteristic equations" (1.5.2). Suppose that Eq. (1.5.3) is to be solved subject to the initial condition

$$R(0, x) = f(x) \tag{1.5.4}$$

Suppose also that the initial value problem

$$u(t) = F(t, 0, u(0), x(0)), \qquad u(0) = f(s)$$
$$x(t) = G(t, 0, u(0), x(0)), \qquad x(0) = s$$

If the functional equations in Theorem 1.5.1 are discrete evolution equations, the proof of the theorem accounts for a single step in the evolution. If no domain and range problems occur, the evolution can be continued by induction.

In order to give an example of a discrete problem and to show the relation to the classical theory of characteristics of Section 1.1, let us consider the discrete system obtained from the differential equations with the standard first-order Euler approximations. Starting with the characteristic equations

$$u' = F(t, u, x), \qquad x' = G(t, u, x)$$

and partial differential equation

$$u_t(t, x) + u_x(t, x)G(t, u, x) = F(t, u, x)$$

we obtain the discrete evolution equations
has a unique solution $\{u = u(t, s), x = x(t, s)\}$ for all $s \in E^n$. Let us further assume that the inverse function $s = s(t, x(t))$ exists. If we set $R(t, x(t)) \equiv u(t, s(t, x(t)))$, we see that

$$R(t, x(t)) \equiv R(t, G(t, 0, R(0, s), s)) = F(t, 0, R(0, s), s).$$

Again, uniqueness of the solution for the initial value problem assures the uniqueness of the solution for (1.5.3) subject to (1.5.4).

$$u(k + 1) = u(k) + \Delta t\, F(k\, \Delta t, u(k), x(k)) \equiv \tilde{F}(k, u(k), x(k))$$
$$x(k + 1) = x(k) + \Delta t\, G(k\, \Delta t, u(k), x(k)) \equiv \tilde{G}(k, u(k), x(k))$$

where $\{k\, \Delta t\}_{k=0}^{N}$ defines a partition of $[0, T]$ and where $u(k)$ is meant to denote $u(k \, \Delta t)$. The related functional equation (1.5.3) assumes the form

$$R(k + 1, x(k) + \Delta t \, G(k \, \Delta t, u(k), x(k))) = R(k, x(k))$$
$$+ \Delta t \, F(k \, \Delta t, R(k, x(k)), x(k))$$

It is readily seen from a Taylor expansion that this equation is an explicit first-order approximation of the continuous partial differential equation. It may be noted that this characteristic theory is easily reformulated for implicit functional evolution equations of the form $F(t, u(t), \tau, u(\tau)) = 0$. However, we shall not pursue this generalization.

Linear difference equations allow a similar simplification of the theory as linear differential equations. Suppose that the system (1.5.2) can be expressed as

$$\begin{aligned} u(k + 1) &= A(k)u(k) + B(k)x(k) + F(k) \\ x(k + 1) &= C(k)u(k) + D(k)x(k) + G(k) \end{aligned} \tag{1.5.5}$$

and that the initial manifold is given by $u(0) = fx(0) + a$ where f is, as usual, an $m \times n$ matrix. If we generate the integral surface for (1.5.3) through this initial manifold by solving the linear difference equations subject to $u(0) = fs + a$, $x(0) = s$ and by eliminating the free vector parameters, we find that u and x are related through the affine transformation

$$u(k) = U(k)x(k) + w(k) \tag{1.5.6}$$

Substitution of this expression into (1.5.3) leads to

$$U(k + 1)[C(k)U(k)x(k) + C(k)w(k) + D(k)x(k) + F(k)] + w(k + 1)$$
$$= A(k)U(k)x(k) + A(k) + w(k) + B(k)x(k) + F(k)$$

This expression has to hold for arbitrary $x(k)$ and hence can be reduced to

$$U(k + 1)[C(k)U(k) + D(k)] - A(k)U(k) - B(k) = 0 \tag{1.5.7}$$
$$w(k + 1) + U(k + 1)[C(k)w(k) + G(k)] - A(k)w(k) - F(k) = 0$$

To satisfy the initial values, we require that

$$U(0) = f, \qquad w(0) = a$$

These equations are the exact analogs of Eqs. (1.2.5) and (1.2.6). In fact, it is not difficult to show that if (1.5.5) represents an Euler approximation to linear differential equations, then (1.5.7) is the Euler approximation to the corresponding Riccati and linear equations.

NOTES

N.1.1. A comprehensive account of the relation between first-order partial differential equations and the associated characteristic ordinary differential equations is given by Courant and Hilbert (1962, II, Chapter 2). A somewhat more elementary treatment is given by Kamke (1950). The approach taken here parallels that of Kamke but develops the theory largely in a coordinate-free setting.

N.1.2. Kamke (1950, p. 321) states this result under the weaker hypothesis $w(x(r)) - u(r) \equiv c$ for an arbitrary constant c but uses $c \equiv 0$ in the proof.

N.1.3. This theorem is essentially due to Kamke (1950) but the proof is made simpler and more general through the use of norm estimates.

N.1.4. It is difficult to establish whether repeated iteration can substantially improve the bound (1.1.8) because the computation is quite messy. An improvement of Theorem 1.1.3 along a different line is possible if the bounds on the derivatives of F and G are defined as functions of t.

N.1.5. Kamke (1950, p. 326) presents this result for the case where u is one-dimensional.

N.3.1. A detailed study containing existence and comparison theorems for the matrix Hermitian Riccati equation as well as an up-to-date bibliography are included in the monograph of Reid (1972). It may be noted that in contrast to Reid's work the stability method of this section does not exploit the relationship with the underlying first-order system of linear equations.

N.3.2. A direct proof of this result may be found for matrices in Kalman and Bucy (1961) and, for Hilbert space equations, in Falb and Kleinman (1966). For a discussion of such infinite-dimensional equations we refer to Chapter 6.

N.3.3. Theorem 1.3.2 is given in Reid (1972) and is interpreted as a dissipative property of Riccati equations. As noted in N. 3.1 our method of proof is different.

N.3.4. As stated, the Hermitian form (1.3.3) remains of value in the more general case where the boundedness of $\langle U(t)x, x \rangle$ implies the boundedness of the spectral radius $\varrho(t)$ of $U(t)$, which in turn assures the boundedness $\| U(t) \|$. Consider, for example, a Riccati equation whose solution is an

$n \times n$ matrix with only nonnegative entries. It follows from the Perron–Frobenius theorem that the spectral radius of $U(t)$ is equal to its largest real positive eigenvalues. Suppose that $\varrho(t) < 1$ on $[0, t_0)$ and $\varrho(t_0) = 1$. Then there exists a nonnegative eigenvector x such that $\| x \| = 1$ and

$$\langle U(t_0)x, x \rangle = 1, \qquad \| U(t_0)x \| = 1.$$

The proof of Theorem 1.3.2 shows that

$$\langle U(t_0)x, x \rangle' = \langle [B(t_0) - C(t_0)]x, x \rangle + \langle [A(t_0) - D(t_0)]x, x \rangle < 0$$

is inconsistent with

$$-\varrho(t) \leq \langle U(t)x, x \rangle \leq \varrho(t) < 1 \qquad \text{on} \quad [0, t_0)$$

It is straightforward to verify that if

$$C(t) = -B(t) \qquad \text{and} \quad D(t) = -A(t)$$

then the negativity of the above Hermitian form is assured whenever $B(t) + A(t)$ have negative row (or column) sums. Thus, our approach also furnishes the global existence theorem of Bellman *et al.* (1965) for certain Riccati equations in transport theory (see also Reid, 1972) where the row sum condition is interpreted as a conservation principle.

Chapter

2

Two-Point Boundary Value Problems

2.1. INITIAL VALUE TECHNIQUES FOR BOUNDARY VALUE PROBLEMS—A SURVEY

Among the various techniques available for the analytical and numerical solution of boundary value problems for differential equations there are a number of methods which attack the given problem by solving instead certain related initial value problems. The best known and most universally applicable initial value method is the so-called shooting method. Because this technique is closely connected with the invariant imbedding approach pursued in this monograph, we shall briefly discuss the method.

Consider a two-point boundary value problem of the general form

$$u' = F(t, u), \qquad f(u(0), u(1)) = 0 \qquad (2.1.1)$$

where $u = (u_1, \ldots, u_m)^T$ and $f = (f_1, \ldots, f_m)^T$. If the equation is integrated subject to the initial value $u(0) = s$, then s has to be chosen such that the solution $u(t, s)$ satisfies the boundary equation

$$f(s, u(1, s)) = 0 \qquad (2.1.2)$$

If the boundary value problem is linear, Eq. (2.1.2) can usually be solved for s in terms of fundamental solutions of (2.1.1). If the equations are nonlinear, an iteration is set up for (2.1.2) to generate a sequence $\{s_n\}$ of initial values, with the hope that $s_n \to s$ and $f(s_n, u(1, s_n)) \to 0$ as $n \to \infty$. This iteration may be Newton's method, successive substitution, or any of the other methods for the solution of nonlinear equations, or, for small systems or simple boundary conditions, it may be a trial and error search method. Of course, for all these methods the solution $u(1, s)$ of the differential equation is required, which usually is available only in numerical form.

The drawback with the shooting method is that the iteration may not converge, particularly if no educated guess is available for s_0. The second problem is of a numerical nature and is already present for linear problems. Equation (2.1.2) may be ill-conditioned and difficult to solve for s due to undesirable exponentially growing solutions of the differential equation. A good deal is known about the shooting method, and we refer to the comprehensive monograph of Keller (1968) for a detailed discussion of the theoretical and practical aspects of the shooting method and its variants for ill-conditioned problems. Related material for second-order equations may be found in the book of Bailey *et al.* (1968). In general, shooting methods and their modifications are extremely powerful for solving boundary value problems if a convergent iteration can be set up.

Different approaches to solving boundary value problems by initial value techniques which, however, appear applicable to linear problems only, involve factoring the differential operators (Taufer, 1966; Vishnevetsky, 1968), the use of projection operators (Guderley and Nicolai, 1966), use of the adjoint and complementary functions (Goodman and Lance, 1956), the method of particular integrals (Miele, 1970), the projective geometry method of Weinel (1965) and the sweep method of Gelfand-Lokutsiyevskii. All of these techniques make essential use of the linear structure of the solution space of a linear differential equation. However, the exact interrelation between these methods has apparently not yet been studied (N.1.1). If nonlinear equations are to be solved, then all techniques must be used in conjunction with a linearization method such as Newton's method.

A noniterative technique for the solution of nonlinear boundary value problems by initial value problems is based on continuous transformation groups (see Na and Hansen (1968) and Belford (1969)). In this method the independent and dependent variables are transformed according to a chosen group of transformations. The parameters of the transformation are then calculated such that the transformed dependent variable satisfies an initial value problem. Once the transformed solution is known, that of the original problem is obtained from the inverse transformation. The advantage of this technique is due to the fact that boundary value problems for nonlinear ordinary differential equations are transformed into initial value problems for ordinary differential equations. However, at this time, the theory is not complete and appears to require considerable analysis of the particular problem to be solved before the applicability of the group theoretic method can be established.

Let us now turn to the method of invariant imbedding which originated in 1957 with a series of papers by Bellman, Kalaba, and Wing (see N.1.2) on radiative transfer and which since then has seen an explosive growth and extensive application. Originally, the derivation of the imbedding equations was inspired by particle transport processes and was based, essentially, on (neutron) particle counting. The term "invariant imbedding" was coined in this context to account for certain invariance properties of the particle stream. Subsequently, the method became divorced from the physical model but its name was retained, and today invariant imbedding is generally considered to be an initial value method for boundary value problems obtained by boundary perturbation techniques. A recent exposition of the method and additional references on invariant imbedding may be found in the book of Lee (1968).

Whereas the computational value of invariant imbedding is well established, the theoretical aspects of the general method are lagging. The subsequent sections are intended to provide a rigorous derivation of the invariant imbedding equations and to show their applicability to wide classes of boundary value problems.

2.2. DERIVATION OF THE INVARIANT IMBEDDING EQUATION

We shall now present a derivation of the invariant imbedding equations from the viewpoint of characteristic theory. For ease of exposition let us first consider a two-point boundary value problem for $(m + n)$-dimensional systems of ordinary differential equations with simple separated boundary conditions that we shall write as

$$
\begin{aligned}
u'(t) &= F(t, u, x), & u(0) &= a \\
x'(t) &= G(t, u, x), & x(T) &= b
\end{aligned}
\tag{2.2.1}
$$

Here $a = (a_1, \ldots, a_m)^{\mathrm{T}}$ and $b = (b_1, \ldots, b_n)^{\mathrm{T}}$ are given vectors; furthermore, for ease of exposition we shall assume that F and G are continuous in t on $[0, T]$ and continuously differentiable in u and x.

As in the shooting method, we shall imbed this boundary value problem into the family of initial value problems

$$u'(t) = F(t, u, x), \qquad u(0) = a$$
$$x'(t) = G(t, u, x), \qquad x(0) = s \tag{2.2.2}$$

where $s = (s_1, \ldots, s_n)^\mathrm{T}$ is an n-dimensional parameter. However, we shall not actually carry out the integration. Instead, we observe from Theorem 1.1.2 that as s ranges over E^n, i.e., as we search for the correct initial value, the resulting solutions $\{u(t, s), x(t, s)\}$, when interpreted as characteristic curves, generate the integral surface $u(t, x)$ of the Cauchy problem

$$u_t + u_x(t, x)G(t, u, x) = F(t, u, x), \qquad u(0, x) = a \tag{2.2.3}$$

where u_x is the Jacobi matrix $(\partial u_i/\partial x_j)$, $i = 1, \ldots, m$; $j = 1, \ldots, n$. Conversely, suppose that the Cauchy problem (2.2.3) has a solution $u(t, x)$ for all x and $t \in [0, T]$. Suppose also that the characteristic $\{u(t), x(t)\}$ subject to $u(T) = u(T, b)$, $x(T) = b$ exists over $[0, T]$. Then by Theorem 1.1.1 this characteristic, which by definition satisfies the differential equations of (2.2.1), must remain imbedded in the integral surface $u(t, x)$ over $[0, T]$. Consequently, $u(0) = u(0, x(0)) = a$. Hence the characteristic $\{u(t), x(t)\}$ through $(u(T, b), b)$ is a solution of the boundary value problem (2.2.1).

The method of invariant imbedding is based on the premise that it is easier to determine $u(T, b)$ from the Cauchy problem (2.2.3) and to integrate (2.2.1) subject to the initial value $(u(T, b), b)$ than to solve the boundary value problem (2.2.1) directly. The system (2.2.3) of m partial differential equations with the same principal part will be called the *invariant imbedding equation* associated with (2.2.1). Its solution through a given initial manifold is indeed frequently obtainable as demonstrated later. For numerical work it may be important to avoid integrating an initial value problem for (2.2.1) because of numerical instabilities (see Example 2.2.10). In this case the following observation is important. Once $u(t, x)$ is known, then $x(t)$ may be found from the lower dimensional system

$$x' = G(t, u(t, x), x), \qquad x(T) = b \tag{2.2.4}$$

This is valid because the solution $\{u(t, x(t)), x(t)\}$ is a characteristic through $(u(T, b), (0))$. Under the smoothness hypothesis on F and G this characteristic must be unique so that $u(t, x(t)) \equiv u(t)$, $x(t) \equiv \hat{x}(t)$, where $\{u(t), \hat{x}(t)\}$ are found by integrating (2.2.1) subject to the initial value $(u(T, b), b)$. Integration of (2.2.3) may be called a forward sweep and that of (2.2.4) the backward (or reverse) sweep. It will be demonstrated later that on occasion Eqs. (2.2.3) and (2.2.4) exhibit a different stability property than Eqs. (2.2.2).

The extension of this approach to more complicated boundary conditions is straightforward. Suppose we are to solve the two-point boundary value problem

$$u'(t) = F(t, u, x), \qquad\qquad u(0) = f(x(0))$$
$$x'(t) = G(t, u, x), \qquad g(u(T), x(T)) = 0$$

(2.2.5)

Its solution, if it exists, is imbedded in the family of initial value problems

$$u'(t) = F(t, u, x), \qquad u(0) = f(s)$$
$$x'(t) = G(t, u, x), \qquad x(0) = s$$

In view of Theorem 1.1.2, the solution is also imbedded in the integral surface $u(t, x)$ of the following initial value problem for the invariant imbedding equation:

$$u_t(t, x) + u_x(t, x)G(t, u, x) = F(t, u, x)$$
$$u(0, x) = f(x)$$

(2.2.6)

Conversely, if $u(t, x)$ exists for all x and for $t \in [0, T]$, and if the equation

$$g(u(T, x), x) = 0 \qquad\qquad (2.2.7)$$

has at least one solution $\hat{x}$, then the characteristic $\{u(t), x(t)\}$ found from the initial value problem

$$u' = F(t, u, x), \qquad u(T) = u(T, \hat{x})$$
$$x' = G(t, u, x), \qquad x(T) = \hat{x}$$

necessarily remains imbedded in $u(t, x)$ and hence is a solution of (2.2.5). Alternatively, if $u(t, x)$ is given, then $x(t)$ may be found from the initial value problem

$$x' = G(t, u(t, x), x), \qquad x(T) = \hat{x} \qquad (2.2.8)$$

We also note that each root of Eq. (2.2.7) defines a distinct solution of (2.2.5). Of course, if (2.2.6) does not have a solution for all x and $t \in [0, T]$, then the invariant imbedding approach may not be valid and other techniques are required to solve (2.2.5). These questions will be considered in Section 2.3.

Equations (2.2.6) and (2.2.7) will prove to be the workhorses for solving boundary value problems with invariant imbedding. For ease of reference we shall summarize the discussion presented so far in the following theorem which at the same time pins down the necessary hypotheses for the system (2.2.6).

THEOREM 2.2.1. Let F and G be continuous in t and locally Lipschitz continuous in u and x. Suppose that the initial value problem (2.2.6) has a solution over the strip $[0, T] \times D$ where D is an open ball in E^n (i.e., $D = \{x: \| x - x_0 \| < d,\ d > 0$ for some $x_0 \in E^n\}$). Assume also that Eq. (2.2.7) has at least one solution $\hat{x} \in D$ and that the base characteristic $x(t)$ through $\hat{x}$ remains in D for all $t \in [0, T]$. Then the characteristic $\{u(t), x(t)\}$ through $(u(T, \hat{x}), \hat{x})$ is a solution of the two-point boundary value problem (2.2.5).

Proof. By definition the characteristic satisfies its characteristic equations, the differential equations of (2.2.5), while the smoothness hypothesis is sufficient to assure that the characteristic remains imbedded in the integral surface. Hence, it satisfies the initial condition $u(0) = f(x(0))$ and therefore is a solution of (2.2.6). ∎

If the boundary conditions at the two endpoints are coupled, we need a slightly different interpretation of the boundary value problem before the characteristic theory approach can be applied. Suppose that the two-point boundary value problem is expressed as

$$y' = H(t, y), \qquad h(y(0), y(T)) = 0 \qquad (2.2.9)$$

where $y = (y_1, \ldots, y_n)^{\mathrm{T}}$ and $h = (h_1, \ldots, h_n)^{\mathrm{T}}$. We shall assume that h is twice continuously differentiable with respect to $y(T)$. Equation (2.2.9) is a general formulation which includes the cases discussed above. In order to solve (2.2.9) we shall introduce two auxiliary functions $r(t)$ and $z(t)$ defined by the equations

$$r(t) = y(0), \qquad z(t) = h(r(t), y(t))$$

By differentiating we can associate (2.2.9) with the additional differential equations and boundary values

$$r' = 0, \qquad\qquad r(0) = y(0)$$
$$z' = h_y(r, y)H(t, y), \qquad z(0) = h(y(0), y(0)),\ z(T) = 0$$

where h_y is the Jacobi matrix $(\partial h_i / \partial y_j)$, and where $z(T) = 0$ is precisely the given boundary condition $h(y(0), y(T)) = 0$.

Collecting these results, we obtain the two-point boundary value problem

$$
\begin{aligned}
z' &= h_y(r, y)H(t, y), & z(0) &= h(y(0), y(0)) \\
r' &= 0, & r(0) &= y(0) & \text{(2.2.10)} \\
y' &= H(t, y), & z(T) &= 0
\end{aligned}
$$

The boundary conditions are now separated, and the above discussion applies so that the missing boundary value $y(T)$ can be found by integrating the $(n + 1)$-dimensional Cauchy problem

$$
\left(\begin{array}{c} \partial z/\partial t \\ \partial r/\partial t \end{array} \right) + \left(\begin{array}{c} \partial z/\partial y \\ \partial r/\partial y \end{array} \right) H(t, y) = \left(\begin{array}{c} h_y(r, y)H(t, y) \\ 0 \end{array} \right)
$$

$$
\left(\begin{array}{c} z(0, y) \\ r(0, y) \end{array} \right) = \left(\begin{array}{c} h(y, y) \\ y \end{array} \right)
$$

and determining the proper $y(T)$ from the system

$$
z(T, y) = 0
$$

The corresponding initial value $y(0)$ is given by $y(0) \equiv r(T, y(T))$. To verify this assertion we note that for given $y(T)$, $r(T)$, and $z(T)$, Eqs. (2.2.10) describe a pure initial value problem. Under the above smoothness hypothesis, it has a unique solution $\{y(t), r(t), z(t)\}$. Let $w(t) = h(r(t), y(t))$, then differentiation shows that $(w - z)' = 0$. Since also $w(T) - z(T) = 0$, it follows that $w(t) = z(t)$ and $z(t) = h(r(t), y(t)) = h(y(0), y(t))$. Hence the partial characteristic $y(t)$ is indeed the solution of (2.2.9).

Let us summarize the results of this section in the following theorem for the general system (2.2.9). For ease of application it will be advantageous to separate those boundary equations involving one point only. Thus, we shall write (2.2.9) in the form

$$
\begin{aligned}
u' &= F(t, u, x, y), & u(0) &= f(x(0), y(0)) \\
x' &= G(t, u, x, y), & g(u(T), x(T), y(T)) &= 0 & \text{(2.2.11)} \\
y' &= H(t, u, x, y), & h(u(0), x(0), y(0), u(T), x(T), y(T)) &= 0
\end{aligned}
$$

With this system we shall associate the functions r, s, and z defined by

$$
r(t) = x(0), \qquad s(t) = y(0)
$$

$$
z(t) = h(f(r(t), s(t)), r(t), s(t), u(t), x(t), y(t))
$$

and yielding the differential equations

$$r' = 0, \qquad s' = 0$$

$$z' = h_u(f, r, s, u, x, y)F(t, u, x, y) + h_x(f, r, s, u, x, y)G(t, u, x, y)$$
$$+ h_y(f, r, s, u, x, y)H(t, u, x, y)$$

with initial values $r(0) = x(0)$, $s(0) = y(0)$, $z(0) = h(f(x(0), y(0)), x(0),$ $y(0), f(x(0), y(0)), x(0), y(0))$. We then can state:

THEOREM 2.2.2. Suppose that the Cauchy problem

$$\begin{pmatrix} \partial u/\partial t \\ \partial r/\partial t \\ \partial s/\partial t \\ \partial z/\partial t \end{pmatrix} + \begin{pmatrix} \partial u/\partial x & \partial u/\partial y \\ \partial r/\partial x & \partial r/\partial y \\ \partial s/\partial x & \partial s/\partial y \\ \partial z/\partial x & \partial z/\partial y \end{pmatrix} (G(t,u,x,y)H(t,u,x,y))^{\mathrm{T}} = \begin{pmatrix} F(t, u, x, y) \\ 0 \\ 0 \\ h_u F + h_x G + h_y H \end{pmatrix}$$

$$\begin{pmatrix} u(0, x, y) \\ r(0, x, y) \\ s(0, x, y) \\ z(0, x, y) \end{pmatrix} = \begin{pmatrix} f(x, y) \\ x \\ y \\ h(f(x, y), x, y, f(x, y), x, y) \end{pmatrix}$$

$$(2.2.12)$$

has a solution for $t \in [0, T]$ and all x and y, and that the equation

$$\begin{pmatrix} g(u(T, x, y)) \\ z(T, x, y) \end{pmatrix} = 0$$

has at least one root $(\hat{x}, \hat{y})$. Then the characteristic $\{u(t), x(t), y(t)\}$ through the point $(u(T, \hat{x}, \hat{y}), \hat{x}, \hat{y})$ is a solution of the boundary value problem (2.2.11).

We note that if the boundary conditions are already separated $(y \equiv H \equiv h = 0)$, then the equation for u is the previously derived imbedding equation. Equation (2.2.12) is the most general invariant imbedding equation for two-point boundary value problems for ordinary differential equations. The same equation was found by Golberg (1971) with a boundary perturbation technique for the case of coupled but additive boundary conditions (N.1.3).

In order to conclude this discussion of invariant imbedding via characteristic theory, let us relate our presentation to some of the terminology and concepts found in the standard literature on the method of invariant imbedding. If the boundary value problem (2.2.1) describes a neutron transport process in a bar of length T, then $u(t)$ and $x(t)$ denote the particle densities of the fluxes to the right and left, respectively, and the solution

$u(t, x)$ of the invariant imbedding equation (2.2.3), when evaluated at $t = T$ and $x = b$, yields the number of particles emerging to the right at $t = T$. The boundary values a and b specify the particles entering the bar at the left and right ends of the bar. Linear boundary conditions of the form $u(0) = fx(0) + a$ indicate an input of fresh particles at $t = 0$ as well as a backscatter of particles moving to the left. The matrix f is known in transport theory as the albedo and $u(t, x)$ is the reflection function. Once the reflection function is known, one can derive an equation for the so-called transmission function which yields the particles emerging to the left at $t = 0$. Thus, the reflection function provides one set of boundary values and the transmission function the remaining boundary values which are consistent with the given boundary values. Let us derive the equation for the transmission function. Suppose that problem (2.2.5) is given and that the solutions $u(t, x)$ of (2.2.6) and $\hat{x}$ of (2.2.7) have been found. We need to determine an x_0 such that the base characteristic defined by

$$x'(t) = G(t, u(t, x), x), \qquad x(0) = x_0 \qquad (2.2.13)$$

satisfies $x(T) = \hat{x}$. Let $r(t)$ be a curve which is constant and equal to $x(0)$ along $x(t)$. Then the solution $r(T)$ of the standard two-point boundary value problem

$$
\begin{aligned}
r' &= 0, & r(0) &= x(0) \\
x' &= G(t, u(t, x), x), & x(T) &= \hat{x}
\end{aligned}
\qquad (2.2.14)
$$

satisfies $r(T) = x_0$. Of course, invariant imbedding is immediately applicable to (2.2.14) so that $r(T) = r(T, \hat{x})$, where $r(t, x)$ is the solution of the initial value problem

$$r_t = r_x G(t, u(t, x), x) = 0, \qquad r(0, x) = x$$

This $r(t, x)$ is the transmission function. We note that it is identical with the function r obtained from Theorem 2.2.2. [For a detailed discussion of the significance of reflection and transmission functions we refer to the monograph of Wing (1962).]

2.3. EXISTENCE THEOREMS FOR TWO-POINT BOUNDARY VALUE PROBLEMS

It is apparent from the preceding section that the existence of the integral surface is intimately connected with existence and uniqueness theorems for boundary value problems defined for its characteristic equations. We shall discuss this connection in some detail to establish the validity of invariant imbedding as a rigorous computational method for certain classes of boundary value problems. Our characteristic theory approach has its advantages and drawbacks for theoretical work. The main advantage is that invariant imbedding provides a ready-made blueprint for proving existence theorems for boundary value problems by (1) establishing the existence of the integral surface through the given initial manifold over a given domain $[0, T] \times D$; (2) demonstrating the existence of a solution $\hat{x} \in D$ of the boundary condition $g(u(T), x(T)) = 0$; and (3) showing that the characteristic through the computed boundary values remains in D. It may be observed that all the theorems of this section follow closely this blueprint. It may also be noted that while the interpretation is different the same general approach is used by Keller (1968) when establishing the existence of solutions with the shooting method. The main drawback is the fact that the existence of a bounded integral surface is not a necessary condition for the existence of a solution for two-point boundary value problems.

Let us begin our discussion of existence theorems by considering first the simple problem (2.2.1).

THEOREM 2.3.1. Suppose that F and G satisfy the conditions of Theorem 1.1.3. Then the boundary value problem

$$u' = F(t, u, x), \qquad u(0) = f(x(0))$$
$$x' = G(t, u, x), \qquad x(T) = b$$

always has a unique solution if $T < \hat{t}$, where $\hat{t}$ is given by (1.1.5) (or (1.1.8)).

Proof. By Theorem 1.1.3 the integral surface $u(t, x)$ of the corresponding imbedding equation exists over $[0, T] \times E^n$, and by Theorem 1.1.1 the characteristic through $(u(T, b), b)$ is a solution of (2.2.1). Uniqueness follows from the uniqueness of the integral surface through $u(0, x) = f(x)$ and the uniqueness of the characteristic through a given point. ∎

A somewhat sharper result is obtained by restricting existence of $u(t, x)$ to a strip as in Theorem 2.2.1. By taking a ball D about the given boundary value $b \in E^n$ and extending the functions F and G outside this ball according to Lemma 1.1.1, we can compute a $\hat{t}$ such that $u(t, x)$ exists over $[0, T] \times D$ provided $T < \hat{t}$. If, moreover, the characteristic through $(u(T, b), b)$ remains in the strip, then (2.2.1) necessarily has a unique solution.

As pointed out in Section 1.1, the estimates (1.1.5) and (1.1.8) for $\hat{t}$ may be quite conservative. On the other hand, Eq. (1.1.11) shows that in specific cases there does indeed exist a $\hat{t}$ beyond which the integral surface cannot be continued. In the invariant imbedding literature $\hat{t}$ is known as the critical length and Theorem 2.3.1 states that the boundary value problem (2.2.1) has a unique solution for all $T \geq 0$ less than the critical length. Furthermore, if $u(T, x)$ exists for all $x \in E^n$, then (2.2.1) has a unique solution for arbitrary $x(T) = x \in E^n$. Thus, $u(t, x)$ is the missing boundary value at t for the two-point problem

$$u' = F(t, u, x), \qquad u(0) = a$$
$$x' = G(t, u, x), \qquad x(t) = x$$

In the usual boundary perturbation approach to invariant imbedding, $u(t, x)$ is interpreted as the boundary value for (2.2.1) corresponding to $u(0) = a$, $x(t) = x$, and the invariant imbedding equation is taken to describe the variation of the boundary value $u(t, x)$ with respect to t and x. With this interpretation, Theorem 2.3.1 is also a statement about the continuity of the solution of (2.2.1) with respect to the boundary values and the interval of integration.

For the more general boundary value problem (2.2.5) we have to assure that $u(t, x)$ exists over $[0, T]$ for all x and that the boundary condition $g(u(T, x), x) = 0$ can be satisfied. To this end the homotopy arguments described in the introduction will be applied. If $u(T, x)$ exists for all x, then two simple homotopies will prove useful. They are, for arbitrary x_0,

$$H_1(\lambda, x) \equiv (1 - \lambda)(x - x_0) + \lambda g(u(T, x), x) = 0 \qquad (2.3.1a)$$
$$H_2(\lambda, x) \equiv g(u(T, x), x) + (\lambda - 1)g(u(T, x_0), x_0) = 0 \qquad (2.3.1b)$$

The corresponding equations for the continuation are

$$dx/d\lambda = -[(1 - \lambda)I + \lambda(g_u u_x + g_x)]^{-1}[g(u(T, x), x) - (x - x_0)] \qquad (2.3.2a)$$
$$dx/d\lambda = -[g_u u_x]^{-1}g(u(T, x_0), x_0) \qquad (2.3.2b)$$

both subject to $x(0) = x_0$. We shall use the homotopy H_1 to prove the following result.

THEOREM 2.3.2. Suppose that F and G satisfy the conditions of Theorem 1.1.3. Suppose further that the boundary condition $g(u(T), x(T)) = 0$ can be expressed as $x(T) = \hat{g}(u(T))$, where $\hat{g}$ has a bounded derivative. Then

(2.2.5) always has a unique solution provided that $T < \hat{t}$ (given by (1.1.5)) and

$$\| \hat{g}_u \| \max\{1, i\} e^{(k+d)T}\{1/(1 - \gamma(T))\} < 1$$

where

$$\gamma(t) = c \max\{1, i\}\{1/(k + d)\}(e^{(k+d)t} - 1)$$

Proof. Let x_0 be arbitrary; then the differential equation (2.3.2a) can be solved over [0, 1] provided $[(1 - \lambda)I + \lambda(g_u u_x + g_x)] = [I + \lambda \hat{g}_u u_x]$ has a bounded inverse. This will certainly be the case if $\| \hat{g}_u u_x \| < 1$ for all $x \in E^n$. By hypothesis $\hat{g}_u$ is bounded. To obtain a bound for u_x, we note that by the chain rule $u_x = u_s s_x$. For u_s we have the estimate $\| u_s \| \leq \max\{1, i\} e^{kt}$ (Theorem 1.1.3). It remains to bound

$$s_x = (x_s)^{-1} = \left\{\psi(t, 0)\left[I + \int_0^t \psi(0, r)G_u u_s \, dr\right]\right\}^{-1}$$

As shown in the proof of Theorem 1.1.3, the estimate

$$\left\| \int_0^t \psi(0, r)G_u u_s \, dr \right\| \leq c \max\{1, i\}\{1/(k + d)\}(e^{(k+d)t} - 1) \equiv \gamma(t) < 1$$

obtains on $[0, \hat{t})$. Hence again by Banach's lemma

$$\| s_x \| = \| x_s^{-1} \| \leq e^{dt}\{1/(1 - \gamma(t))\}$$

Combining these results, we see that we need

$$\| \hat{g}_u u_x \| \leq \| \hat{g}_u \| \max\{1, i\} e^{(k+d)T}\{1/(1 - \gamma(T))\} < 1 \quad \blacksquare$$

(Note that this inequality implies that $x = \hat{g}(u(T, x))$ is a contraction on E^n so that the theorem can also be found without homotopy arguments.)

The same argument may be used in conjunction with the bound (1.1.8). A straightforward calculation shows that we require $T < \hat{t}$ (given by (1.1.8)) and

$$\| \hat{g}_u \| [i + \max\{1, i\}(b/k)(e^{kT} - 1)]e^{(d+a)t}\{1/(1 - \gamma(T))\} < 1$$

where now

$$\gamma(t) = \left[\frac{ci}{a + d} + \max\{1, i\} \frac{cb}{k(k + d + a)}\right][e^{(k+d+a)t} - 1]$$

It follows from Theorem 2.3.2 that invariant imbedding is applicable whenever $\| \hat{g}_u \| i < 1$ and T is sufficiently small. In the limit as $T \to 0$ this requires that $x = \hat{g}(f(x))$ is a contraction mapping which is an acceptable asymptotic behavior.

In many applications the boundary condition $g(u(T), x(T)) = 0$ cannot be solved for $x(T)$. In this case the homotopy (2.3.1b) is useful. For definiteness let us suppose that $g(u(T), x(T)) \equiv u(T) - b = 0$. We can prove the following result.

THEOREM 2.3.3. Suppose that $i \geq \| f_x y \| / \| y \| \geq j$ and that F and G satisfy the conditions of Theorem 1.1.3. Then the boundary value problem (2.2.5) with $g(u(T), x(T)) \equiv u(T) - b = 0$ has a unique solution provided $t < \min\{\hat{\imath}, \hat{t}\}$, where $\hat{\imath}$ is given by (1.1.5) and

$$\hat{t} = 1/(k + a) \ln(1 + j(k + a)(b \max\{1, i\}))$$

Proof. We observe that the differential equation (2.3.2b) will have a unique solution provided $g_u u_x \equiv u_x(T, x) = u_s(T, x)s_x(T, x)$ has a uniformly bounded inverse. The proof now becomes similar to that of Theorem 1.1.3. Using the same notation, we can write

$$u_s(t) = \phi(t, 0)f_s\left[1 + \int_0^t f_s^{-1}\phi(0, r)F_x x_s \, dr\right]$$

Hence, u_s is nonsingular provided $\| \int_0^t f_s^{-1}\phi(0, r)F_x x_s \, dr \| < 1$. Since $\| x_s \| \leq \max\{1, i\}e^{kt}$, it follows that this condition is always satisfied if t chosen such that $t < \hat{t}$ where $\hat{t}$ is a solution of the equation

$$(b/j) \max\{1, i\}\{1/(k + a)\}(\exp[(k + a)\hat{t}] - 1) = 1$$

Since for given $T < \min\{\hat{\imath}, \hat{t}\}$ the matrix s_x is nonsingular, it follows that u_x is nonsingular. ∎

Theorem 2.3.3 applies to the boundary value problem (2.2.11) after adjoining the functions r, s, and z. For particular problems somewhat more precise results are available. Consider, for example, the problem

$$y' = H(t, y), \qquad Ay(0) + By(T) - b = 0$$

If $z(t) = Ay(0) + By(t) - b$ then the invariant imbedding formulation for the two-point boundary value problem

$$z' = By'(t) = BH(t, y), \qquad z(0) = (A + B)y(0) + b$$
$$y' = H(t, y), \qquad z(T) = 0$$

is given as

$$\partial z/\partial t + (\partial z/\partial y)H(t, y) = BH(t, y)$$
$$z(0, y) = (A + B)y - b$$

We proceed as in the proof of Theorem 2.3.3. If $\| H_y(t, y) \| \leq C(t)$ for all $y \in E^n$ where $C(t)$ is integrable, we obtain the bound

$$\| y_s(t) - I \| \leq \exp\left(\int_0^t C(s)\, ds \right) - 1$$

Hence $z_s(t) = A + By_s(t) = A + B + B(y_s(t) - I)$ is invertible provided $(A + B)$ is nonsingular and

$$\| (A + B)^{-1}B \| \left(\exp \int_0^t C(s)\, ds - 1 \right) < 1$$

Since y_s necessarily is invertible, it follows that $z_y = z_s s_y$ is invertible so that $z(T, y) = 0$ has a unique solution. This is the same result derived by Keller (1968, p. 16) with the shooting method.

Some improvement of Theorems 2.3.2 and 2.3.3 is possible through iterating on the bounds for $\| u_s \|$ and $\| x_s \|$ as outlined following Theorem 1.1.3, or through the use of Lemma 1.1.1; however, they will rarely lose their local character. Stronger results are obtainable if the boundary value problem has the proper structure which allows solving the Riccati equation (1.4.3). For example, we can state:

THEOREM 2.3.4. Suppose that the boundary value problem

$$u' = F(t, u, x), \qquad u(0) = f(x(0))$$
$$x' = G(t, u, x), \qquad u(T) + g(x(T)) = 0 \tag{2.3.3}$$

has bounded solutions for all initial values $(f(s), s)$. Suppose further that f_x, g_x, F_x, and G_u are positive definite and that $F_u - G_x^{\mathrm{T}} = \alpha I$ for a scalar α uniformly in u and x. Then the boundary value problem (2.3.3) has a unique solution.

Proof. We shall use the homotopy (2.3.1a) for an arbitrary x_0. We need to prove that $[(1 - \lambda)I + \lambda(g_u u_x + g_x)] = [(1 - \lambda)I + \lambda u_x + \lambda g_x]$ has a bounded inverse. It is readily verified that under the hypotheses the derivative u_x, obtained from the Riccati equation (1.4.3) exists and is positive definite on $(0, T]$. Hence $[(1 - \lambda)I + \lambda u_x + \lambda g_x]$ is positive definite for $\lambda \in [0, 1]$ and the result follows. ∎

If we apply Theorem 2.3.4 to two scalar equations, then the conditions of the theorem require, beyond existence of solutions for the initial value problem, only that $F_x \geq 0$, $G_u \geq 0$ and $f_x \geq 0$, $g_x \geq 0$. In fact, theorem 2.3.4 may be viewed as an extension of an existence theorem of Keller (1968, p. 9) for the problem

$$u'' = G(t, u, u')$$

$$u(0) = \alpha u'(0) + \beta, \qquad \alpha \geq 0$$

$$u(T) + \gamma u'(T) = \delta, \qquad \gamma \geq 0$$

We shall assume that G is uniformly Lipschitz continuous in u, u', that $\partial G/\partial u \geq 0$ and that $|\,\partial G/\partial u'\,|$ is uniformly bounded. If we identify u' with x and work with the corresponding first-order system, then the characteristic equations have solutions for arbitrary initial values and invariant imbedding assures that the proper initial value at $t = T$ can be found from the equation $u(T, x) + \gamma x = \delta$. Here $u(t, x)$ is the solution of the initial value problem

$$\partial u/\partial t + (\partial u/\partial x)G(t, u, x) = x, \qquad u(0, x) = \alpha x + \beta$$

The proof of Theorem 2.3.4 assures that $u(t, x)$ exists for $t \geq 0$ and all x and that $u_x(t, x)$ is positive and uniformly bounded. Hence the equation $g(x) \equiv u(T, x) + \gamma x - \delta = 0$ has a unique solution $\hat{x}$ and the characteristic $\{u(t), x(t)\}$ through $(u(T, \hat{x}), \hat{x})$ is the solution of the given boundary value problem. For systems of equations the theorem is most useful when reducing second-order systems of the form

$$u'' = H(t, u)$$

to first-order form. The conditions on F and G in the theorem are satisfied if H_u is positive definite. Of course, Theorem 2.3.4 is particularly applicable to linear boundary value problems because the integral surface $u(t, x)$ can be expressed as $u(t, x) = U(t)x + v(t)$, where U is the solution of the Riccati equation (1.2.5) and v that of Eq. (1.2.6). To consider a meaningful linear example we shall work with the following self-adjoint $2n$th-order two-point boundary value problem

$$Ly(t) = g(t), \qquad y^{(k)}(0) = y^{(k)}(T) = 0, \quad 0 \leq k \leq n - 1 \qquad (2.3.4)$$

where L is the differential operator

$$Ly(t) = \sum_{j=}^{n} (-1)^{j+1}(p_j(t)y^{(j)})^{(j)}, \qquad n \geq 1$$

If the coefficients p_j are j times continuously differentiable then the $2n$th-order equation can be reduced to the $2n \times 2n$ first-order system (Coddington and Levinson, 1955, p. 206)

$$\begin{pmatrix} 0 & I \\ -I & 0 \end{pmatrix} \begin{pmatrix} u \\ x \end{pmatrix}' + \begin{pmatrix} C(t) & A(t) \\ A^T(t) & B(t) \end{pmatrix} \begin{pmatrix} u \\ x \end{pmatrix} = \begin{pmatrix} G(t) \\ 0 \end{pmatrix}$$

where $u = (u_1, \ldots, u_n)^T = (y^{(0)}, \ldots, y^{(n-1)})^T$ and where the components of $x = (x_1, \ldots, x_n)$ are given by

$$x_j = -p_j(t)y^{(j)} + (p_{j+1}(t)y^{(j+1)})' - \cdots + (-1)^{n+j+1}(p_n(t)y^{(n)})^{(n-1)}$$

G is the vector $G(t) = (g(t), 0, \ldots, 0)^T$ and the $n \times n$ matrices are given by

$$(A(t))_{ij} = \delta_{(i-1)j}, \quad (B(t))_{ij} = -1/p_n(t) \, \delta_{in} \, \delta_{nj}, \quad (C(t))_{ij} = p_{i-1}(t) \, \delta_{ij}.$$

Writing out the first-order system we obtain

$$u' = A^T u + B(t)x$$
$$x' = -C(t)u - A^T x + G(t)$$
$$u(0) = u(T) = 0.$$

From Eq. (1.2.5) we see that the corresponding Riccati equation is

$$U' = B(t) + A^T U + UA + UC(t)U, \qquad U(0) = 0$$

Since both B and C are diagonal matrices this is a Hermitian equation to which the theory of Section 1.3 applies. In particular, we can prove

THEOREM 2.3.5. If p_n and p_0 are strictly positive and p_j is nonnegative for $[0, T]$ for $1 \leq j \leq n - 1$, then the first-order system corresponding to the Dirichlet problem (2.3.4) can be solved with invariant imbedding.

Proof. Since B is negative and C nonnegative semidefinite it follows from Corollaries 1.3.1 and 1.3.2 that the Riccati equation has a bounded negative semidefinite global solution on $[0, T]$. It remains to be shown that we can compute a consistent boundary value $\hat{x}$ from the boundary condition at T which in this case assumes the form

$$u(T) = 0 = U(T)x + v(t)$$

This will be possible if $U(T)$ is invertible. We shall give an indirect proof of this assertion. It follows from

$$\int_0^T y(t)Ly(t)\,dt = \sum_{j=0}^n \int_0^T p_j(t)y^{(j)}(t)^2\,dt$$

and the positivity of p_0 that the associated homogeneous Dirichlet problem

$$Ly(t) = 0, \qquad y^{(k)}(0) = y^{(k)}(T) = 0, \quad 0 \le k \le n-1$$

has the unique solution $y(t) \equiv 0$. If we apply invariant imbedding to the homogeneous problem the same Riccati equation results while $v(t) \equiv 0$. If $U(T)$ were singular there would exist a nonzero solution $\hat{x}$ of $U(T)x = 0$ which would lead to a nonzero solution of the Dirichlet problem for $Ly(t) = 0$. Since this cannot be, the matrix $U(T)$ must be nonsingular and the theorem is proved. ∎

It may be noted from the phrasing of Theorem 2.3.5 that the emphasis was on the applicability of the initial value method rather than the existence of a solution. It is well known that stronger results can be proven for (2.3.4) by exploiting its variational structure (Ciarlet *et al.*, 1968). If this structure is lost and a reduction to a formally self-adjoint system no longer is possible, the corresponding Riccati equation is not Hermitian. In this case Theorem 1.3.2 can prove useful. As a hypothetical illustration consider the third-order problem

$$y''' = a(t)y + b(t)y' + d(t)y'' + g(t)$$
$$y(0) = y'(0) = y''(T) = 0$$

From the usual reduction $u_1 = y$, $u_2 = y'$, $u_3 = y''$ we obtain the first-order system

$$u' = A(t)u + B(t)x$$
$$x' = C(t)u + D(t)x - g(t)$$

where

$$A = \begin{pmatrix} 0 & 1 \\ 0 & 0 \end{pmatrix}, \quad B = \begin{pmatrix} 0 \\ 1 \end{pmatrix}, \quad C(t) = (a(t)b(t)), \quad \text{and} \quad D(t) = d(t)$$

The associated Riccati equation is

$$U' = B + AU - d(t)U - UC(t)U, \quad U(0) = 0$$

where U is a 2×1 matrix. By Theorem 1.3.2 it will have a bounded solution on $[0, T]$ provided

$$\Omega = 2\{\langle (B(t) - C^T(t))y, x\rangle + \langle Ax, x\rangle - d(t)\} < 0$$

or

$$\Omega = 2(-a(t)yx_1 + (1 - b(t))yx_2 + x_1x_2 - d(t)) < 0$$

Since $\| x \| = | y | = 1$ we can use the estimate $| x_1x_2 | \leq \frac{1}{2}$ to obtain

$$\Omega/2 > | a(t) | + | b(t) | + \tfrac{3}{2} - d(t)$$

so that $\Omega < 0$ if

$$d(t) > | a(t) | + | b(t) | + \tfrac{3}{2}$$

Hence the Riccati equation has a bounded solution $U(t)$ whenever $d(t)$ is sufficiently large. The initial value at $t = T$ consistent with the given data is $(v_1(T), v_2(T), 0)$ where v is the solution of the linear equation (1.2.6).

Finally, we note that the scalar problem

$$u'' = -u, \qquad u(0) = u(2) = 0$$

has the unique solution $u(t) \equiv 0$; invariant imbedding applied to the equivalent first-order system

$$u' = x, \qquad x' = -u$$

leads to the Riccati equation

$$U' = 1 + U^2, \qquad U(0) = 0$$

which has the solution $U(t) = \tan t$. Since $U(t) \to \infty$ as $t \to \pi/2$ it follows that invariant imbedding is not applicable to this problem.

2.4. THE NUMERICAL SOLUTION OF NONLINEAR TWO-POINT BOUNDARY VALUE PROBLEMS

In this monograph invariant imbedding is primarily intended to furnish numerical methods for the solution of boundary value problems. Whether a numerical scheme based on invariant imbedding or on alternate methods such as finite difference, Galerkin, or shooting techniques, coupled with a suitable iteration should be applied will depend partly on the structure of the problem and largely on the inclination of the user. In fact, use of a particular solution technique is often determined more by its familiarity or availability to the user than its mathematical suitability. (This, incidentally, was the reason why for many of the calculations below a fourth-order Runge–Kutta integrator was used instead of possibly better suited interpolation or predictor corrector schemes.) It is hoped that the numerical examples of the next sections convey the author's belief that invariant imbedding is a useful, straightforward (i.e., mechanically) to apply, and readily computer coded, usually noniterative method for a diversity of free and fixed, linear and nonlinear boundary value problems.

We shall start the discussion of invariant imbedding as a numerical technique by considering the two-point problem

$$
\begin{array}{ll}
u' = F(t, u, x), & u(0) = f(x(0)) \\
x' = G(t, u, x), & g(u(T), x(T)) = 0
\end{array}
\tag{2.4.1}
$$

where $u = (u_1, \ldots, u_m)^{\mathrm{T}}$ and $x = (x_1, \ldots, x_n)^{\mathrm{T}}$, and where F and G are m- and n-dimensional functions.

As we have seen in Section 2.2, problems with coupled boundary conditions of the form $h(u(0), x(0), u(T), x(T)) = 0$ can be cast into the form (2.4.1) by adjoining auxiliary functions and, if necessary, by renaming u and x. Hence, problem (2.4.1) seems sufficiently general.

Throughout the discussion of invariant imbedding as a numerical tool we shall assume that the functions F and G are sufficiently well behaved to permit the operations to which they will be subjected. Continuity in t and differentiability in u and x are usually required. This is not meant to imply that the method will not work when singularities are present. Indeed, Example 3.3.2 demonstrates that invariant imbedding may only be minimally affected by such singularities. In addition we shall always suppose that the interval $[0, T]$ is finite. From a numerical point of view this is not too serious since infinite intervals are commonly truncated at some large but finite T.

We shall now translate Theorem 2.2.1 into an explicit algorithm which is commonly called the method of invariant imbedding.

Algorithm. The Method of Invariant Imbedding for Nonlinear Two-Point Boundary Value Problems

The boundary value problem

$$
\begin{array}{ll}
u' = F(t, u, x), & u(0) = f(x(0)) \\
x' = G(t, u, x), & g(u(T), x(T)) = 0
\end{array}
\tag{2.4.1}
$$

can be solved in the following three steps:

Step 1. Integrate the initial value problem

$$\frac{\partial u}{\partial t}(t, x) + \frac{\partial u}{\partial x}(t, x)G(t, u, x) = F(t, u, x), \ u(0, x) = f(x) \quad (2.4.2)$$

where in the notation of Section 0.2 the operator $\partial u/\partial x$ (sometimes also written u_x) stands for the $m \times n$ Jacobian matrix

$$\partial u/\partial x \equiv (\partial u_i/\partial x_j) \qquad i = 1, \ldots, m, \quad j = 1, \ldots, n$$

Step 2. Find a solution $\hat{x}$ of the equation

$$g(u(T, x), x) = 0 \quad (2.4.3)$$

Step 3. Integrate the initial value problem

$$x' = G(t, u(t, x), x), \qquad x(T) = \hat{x} \quad (2.4.4)$$

The functions $\{u(t) \equiv u(t, x(t)), x(t)\}$ are a solution of the boundary value problem (2.4.1).

It may not be practical to retain the solution $u(t, x)$ of (2.4.2) over a sufficiently large domain to allow integration of (2.4.4). In this case, Step 3 can be replaced by

Step 3'. Integrate the initial value problem

$$\begin{aligned} u' &= F(t, u, x), & u(T) &= u(T, \hat{x}) \\ x' &= G(t, u, x), & x(T) &= \hat{x} \end{aligned} \quad (2.4.5)$$

At first sight the above solution algorithm may look rather complicated. While concerned with ordinary differential equations, we end up with a partial differential equation. But we have gained a major simplification. The partial differential equation (2.4.2), the so-called invariant imbedding equation corresponding to the boundary value problem (2.4.1), is of evolution type and subject to a given initial value so that, at least in principle, its solution at all times is completely determined by the initial value (or manifold) $u(0, x) = f(x)$. Thus, explicit or implicit marching techniques can be used for its numerical solution. The boundary condition $g(u(T), x(T)) = 0$ does not come into play until after the invariant imbedding equation has been integrated. The same advantage is present in the shooting method and stands in contrast to finite difference methods for (2.4.1) where the boundary condition at $t = T$ is part of the resulting nonlinear algebraic system. Once $u(T, x)$ has been computed and the boundary value $\hat{x}$ has been determined from Step 2, we are again faced with an initial value problem in Step 3 or 3′, but this time for ordinary differential equations. There is considerable help available for solving the latter problem. In summary, the boundary value problem (2.4.1) has been converted into a sequence of initial value problems.

Setting up the invariant imbedding equations involves several choices which can materially influence the complexity of the resulting algorithm. Given a two-point boundary value problem like $y' = H(t, y)$ for an n-dimensional vector y ($n \geq 2$), it is generally left to the user of invariant imbedding which components of y to identify with the so-called surface characteristic u and the base characteristic x.

In order to carry out the first step of our algorithm, an initial value of the type $u(t_0) = f(x(t_0))$ is required at some point t_0 of $[0, T]$. In many problems the boundary conditions are given as $u(0) = f(x(0))$, $x(T) = g(u(T))$. In this case, neither equation in (2.4.1) is singled out as surface or base characteristic equation and a choice will have to be made, either on the basis of the desired final result or on the structure of the functions F and G. For example, in the first case the boundary values may be given as $u(0) = a$, $x(T) = b$ and only the boundary value $x(0)$ is of interest. It now is reasonable to consider $x(t)$ as surface and $u(t)$ as base characteristic and to integrate backward the invariant imbedding equation

$$\partial x/\partial t + (\partial x/\partial u)F(t, u, x) = G(t, u, x), \qquad x(T, u) = b$$

The final answer is obtained by evaluating $x(0) = x(0, a)$. In the second case it may be advantageous from a numerical point of view to identify $F(t, u, x)$ with the more complicated right-hand side in (2.4.1) to reduce the nonlinearity of the resulting imbedding equation. Another option is choosing the more complicated boundary function f or g as initial manifold because the numerical integration of the imbedding equation will generally not be influenced by the initial manifold, while the boundary condition in Step 2 can produce thorny problems. In what follows it will be assumed that the choice has been made and that u denotes the surface and x the base characteristic. Exceptions to this convention will be clearly noted.

Theorem 2.2.1 assures that the *exact* solution obtained from the method of invariant imbedding is the *exact* solution of the boundary value problem (2.4.1). However, it was already pointed out in Section 1.4, that the initial value problem (2.4.2) rarely will admit an exact (closed form) solution. Hence, we are faced with integrating (2.4.2) with a suitable numerical scheme and then going through Steps 2 and 3 with already approximate input. Let us discuss execution of Steps 1, 2, and 3 in some detail.

Equation (2.4.2) is a quasi-linear first-order hyperbolic system (with real characteristics), and equations of this type have received some consideration in the literature. We shall outline several techniques for solving (2.4.2) that appear to merit consideration for the solution of practical problems. All of these methods are based on a finite difference formulation. This necessarily restricts integration of (2.4.2) over a finite set, usually a "strip" or "triangle," in E^n. The shape of the proper domain is often determined by the boundary condition (2.4.3). If g can be written as $x(T) = b$, we may get by with a "triangle" with apex at b and base on the initial manifold $u(0, x) = f(x)$.

For a more complicated function g we may need a wide strip in which to search for $\hat{x}$. In addition, once $\hat{x}$ is known, we have to assure that the characteristic $\{u(t), x(t)\}$ through $(u(T, \hat{x}), \hat{x})$ does not leave the "strip" or triangle. As Example 2.5.3 shows, failure to do so can lead to invalid numerical solutions. Some *a priori* knowledge about the behavior of the solution of the boundary value problem can help considerably when placing the mesh on the (t, x)-plane initially.

The simplest scheme for the numerical solution of the invariant imbedding equation is the explicit finite difference method of Courant *et al.* (1952) [see also Forsythe and Wasow (1960, p. 49)]. This method consists of placing a mesh on the (t, x)-plane, replacing u_t by its forward finite difference and u_x by a difference quotient at the previous time level (N.4.1). Formally, either a forward or a backward divided difference can be used; the choice of the correct difference expression is dictated by geometric considerations as described in detail in Forsythe and Wasow. Briefly, if at the mesh point (t, x) the ith base characteristic equation

$$dx_i/dt = G_i(t, u, x)$$

defines a positive slope, then the derivatives $\partial u_j/\partial x_i$, $j = 1, \ldots, m$ at the previous time level $t - \Delta t$ must be replaced by backward quotients

$$\partial u_j/\partial x_i = \{u_j(t - \Delta t, x_i) - u_j(t - \Delta t, x_i - \Delta x_i)\}/\Delta x$$

because $x_i(t - \Delta t)$ will pass between x_i and $x_i - \Delta x_i$. For a scalar equation in t and x, the explicit method leads to the following formula for advancing the solution $u(t, x)$ from time level to time level.

$$u(t_{i+1}, x_j) = u(t_i, x_j) - (\Delta t/\Delta x)[u(t_i, y) - u(t_i, z)]G(t_i, u(t_i, x_j), x_j)$$
$$+ \Delta t\, F(t_i, u(t_i, x_j), x_j) \tag{2.4.6}$$

where $y = x_j$, $z = x_j - \Delta x$ if $G(t_i, u(t_i, x_j), x_j) \geq 0$ and $y = x_j + \Delta x$, $z = x_j$ if $G(t_i, u(t_i, x_j), x_j) \leq 0$. This explicit method is known to be convergent but requires a constraint on the ratio of $\Delta t/\Delta x$ in order to assure stability. For a recent study of the explicit method for systems we refer to the paper of Shampine and Thompson (1970).

Frequently in applications the base characteristic $x(t)$ through $x(0) = 0$ has the solution $x(t) \equiv 0$. Consequently, the characteristic curve $u(t)$ is readily obtainable (possibly numerically) over this base curve from

$$u' = F(t, u, 0), \quad u(0) = f(0)$$

In this case the semi-implicit scheme of Shampine and Thompson (1970) can be used. For ease of exposition let us deal first with scalar equations. According to this method, we have to solve the equations

$$\frac{u(t_{i+1}, x_j) - u(t_i, x_j)}{\Delta t} + \frac{u(t_{i+1}, x_j) - u(t_{i+1}, x_{j-1})}{\Delta x} G(t_i, u(t_i, x_j), x_j)$$
$$= F(t_i, u(t_i, x_j), x_j) \tag{2.4.7}$$

Since $u(t_{i+1}, 0)$ is given, equation this can be solved explicitly for $u(t_{i+1}, x_j)$, $j = 1, 2, \ldots$. This method is also convergent; in addition, it is unconditionally stable if the function G is nonnegative.

For ease of reference let us write the analog of Eq. (2.4.7) for systems where u and x are m- and n-dimensional. Using the notation of Shampine and Thompson we can write

$$\frac{u_l(t_{i+1}, x) - u_l(t_i, x)}{\Delta t} + \sum_{k=1}^{n} \frac{u_l(t_{i+1}, x) - u_l(t_{i+1}, B_k x)}{\Delta x} G_k(t_i, u(t_i, x), x)$$
$$= F_l(t_i, u(t_i, x), x), \quad l = 1, \ldots, m$$

where x is a mesh point $(x_{j_1}^1, \ldots, x_{j_n}^n)$ and where $B_k x$ is defined as the shift operator

$$B_k x = (x_{j_1}^1, \ldots, x_{j_k}^k - \Delta x, \ldots, x_{j_n}^n), \quad k = 1, \ldots, n$$

This difference equation can also be solved explicitly. Moreover, it is unconditionally stable if all components of G are nonnegative. If this is known *a priori* for the nonlinear function G, then negative computed values of G_k (due to the computed rather than exact solution $u(t_i, x)$) should be replaced by zero to improve accuracy and stability of the algorithm.

A third finite difference method is the conservation scheme of Lax (1954) [see also Forsythe and Wasow (1960, p. 84)]. For scalar equations the finite difference analog of Eq. (2.4.2) is written in averaged form as

$$\frac{1}{\varDelta t} \left[u(t + \varDelta t, x) - \frac{u(t, x + \varDelta x) + u(t, x - \varDelta x)}{2} \right]$$
$$+ \frac{u(t, x + \varDelta x) - u(t, x - \varDelta x)}{2\varDelta x} \, G(t, u(t, x), x) = F(t, u(t, x), x)$$

$$(2.4.8)$$

This method is also subject to restrictions on the ratio of $\varDelta t/\varDelta x$. In general, this method appears to be not well understood theoretically but attractive numerically (see N.4.2) when integration over a triangle is sufficient.

For multidimensional base characteristics $(x = (x_1, \ldots, x_n))$, $n > 1$, all of the difference methods proposed for Step 1 reflect the problem inherent in the invariant imbedding formulation, namely, we have to integrate over an n-dimensional strip or triangle; hence, even for small systems the number of mesh points will quickly exhaust computer storage and calculation capacity. Thus x can have only few components. This constraint must also be kept in mind when choosing the base and surface characteristics.

A different method for hyperbolic systems like (2.4.2), also shown to be convergent by Courant *et al.* (1952), is based on integrating along the characteristic directions. For differential equations with the same principal part constructing $u(t, x)$ from the characteristics will amount to the shooting method. We shall see examples later on, where this approach will fail because the characteristic equations admit unstable solutions. If this is not the case, then shooting and the concept of invariant imbedding (i.e., of the generation of an integral surface) can possibly be combined in the obvious manner. We integrate the characteristic equations for a selected finite range of initial values $\{s_i\}$. A smooth surface $u(T, x)$ is then generated which passes through the computed values $\{u(T, s_i)\}$. This surface is used for computing the missing boundary value from $g(u(T, x), x) = 0$.

Step 2 in the method of invariant imbedding is difficult to discuss without reference to particular problems. If $g(u(T), x(T)) \equiv x(T) - b = 0$, then we obtain $u(T)$ simply by evaluating $u(T, b)$ so that invariant imbedding is a direct (noniterative) method. Nonlinear functions g are considerably more difficult to handle. For scalar equations and values of $u(T, x)$ at discrete mesh points, an interpolation and search technique is easy to use and generally of sufficient accuracy when compared to the integration of (2.4.2). Multidimensional systems probably will require *ad hoc* methods for the solution of (2.4.3).

In this respect it is particularly troublesome that $u(T, x)$ is given only at the mesh points x_i. It is conceivable that $g(u(T, x_i), x_i) \neq 0$ as i runs over all mesh points. Hence we need interpolation schemes to complete a surface $u(T, x)$ through the data points, a difficult problem for large n, and solve $g(u(T, x), x) = 0$ which for strongly nonlinear g may in itself be a formidable task. Alternatively, one may look for the coordinate x_i which minimizes $| g(u(T, x_i), x_i) |$. Both approaches may be costly in terms of computer time and questionable analytically. However, once $\hat{x}$ is known, the boundary value problem (2.4.1) may be considered solved because standard numerical integration techniques for ordinary differential equations are applicable to (2.4.4) or (2.4.5).

Whenever a numerical method is needed to execute Steps 1–3 of the method of invariant imbedding, it would, of course, be preferable to choose a technique which is convergent, stable, and has a high order of convergence. Unfortunately, in practice the structure of F and G often preclude verification of the hypotheses required for convergence and stability, so that one can do little more than select a numerical method and observe the behavior of the computed results as the mesh sizes are varied. Numerical experiments appear to indicate that the more implicit the finite difference analog for (2.4.2), the more reliable the results are. A number of variations of the expressions (2.4.6)–(2.4.8) come to mind in this context, none of which come equipped with a theoretical foundation. Fortunately, once $\hat{x}$ has been found, the integration of (2.4.5) can generally be carried out with a high degree of accuracy and a check of the boundary condition

$$u(0) = f(x(0))$$

will provide some indication of the quality of the computed boundary value $\hat{x}$.

So far we have side-stepped the question of order of convergence for the proposed finite difference methods. It is apparent from the formulas (2.4.6)–(2.4.8) that only first-order difference approximations are used so that the order of convergence can be expected to be at most linear in Δt and Δx. Precise *a priori* error estimates are given for the explicit and semi implicit method by Shampine and Thompson (1970) provided the data functions satisfy certain sign requirements which allow the application of a maximum principle.

If it is possible to compare the solutions of boundary value problems for the same differential equations but different boundary values, the back integration of the characteristic equations (Step 3' in the method of invariant imbedding) lends itself for an *a posteriori* error bound which may be thought of as quantifying the intuitive feeling usually present when checking how well the computed solutions satisfy the boundary condition $u(0) = f(x(0))$. Specific results are given for certain linear systems in Section 2.7. For general nonlinear boundary value problems such comparison theorems are not available. In this case the computed results may be considered acceptable whenever substitution of the computed boundary values into the prescribed boundary conditions produces "physically tolerable" errors. In the examples given below the computation was terminated whenever this criterion was met. In all our work a check of the computed results via (2.4.5) was found to be indispensable for nonlinear problems. It involved little effort and gave confidence in the computed results.

There exists an alternative approach to solving the boundary value problem (2.4.1) by invariant imbedding. One can approximate the differential equations by standard difference methods derived, for example, from Euler's method, the trapezoidal rule, the Runge–Kutta method, etc. In this way we obtain a two-point boundary value problem for a system of difference equations to which the theory of Section 1.5 can be applied. This approach leads to some of the numerical schemes obtained by Lee (1968) with boundary perturbation techniques. One may conjecture that any numerical scheme for (2.4.2) corresponds to the exact invariant imbedding equation (1.5.3) for a difference formulation derived from (2.4.1). We shall pursue this topic in Section 2.11.

Finally, it should be pointed out that for particular problems, it may well be advisable to solve the given nonlinear boundary value problem iteratively with quasi-linearization. Invariant imbedding can now be used to find a reasonable starting value for the iteration by solving the nonlinear problem, and to solve the linear boundary value problem of each iteration. This approach, first pointed out by Bellman and Kalaba (1964), is discussed in detail by Lee (1968, Chapter 7). Additional comments and examples may be found in the paper of Allen *et al.* (1969). Although not explored in this text, linearization and the use of invariant imbedding is especially worth considering when the base characteristic has many components since linear problems are not as sensitive to the dimensions of u and x (see Section 2.7).

A priori no one method appears preferable for all problems.

2.5. NUMERICAL EXAMPLES

Some technical problems will be considered which lead to two-point boundary value problems. Our discussion is intended not only to illustrate the applicability of invariant imbedding for realistic problems, but also to sound a note of caution against its indiscriminate use.

Example 2.5.1. Chemical Reaction Problems

The Arrhenius-type steady-state reaction equations for a hydrocracker used in oil refining can be expressed in the following general form:

$$du/d\xi = g(u, y)(\alpha y - \beta), \qquad u(0) = \hat{u}$$
$$dy/d\xi = -g(u, y), \qquad\qquad y(1) = \hat{y} \tag{2.5.1}$$

where $g(u, y) = \varepsilon y \exp(-\gamma/(u + \delta))$, and where all Greek letters denote positive constants. Here, u designates the temperature along the normalized reactor length ξ, while y is the carbon–carbon bond remaining in the oil during hydrogen quenching. To optimize operation of the unit, the inlet temperature $u(1)$ has to be found (N.5.1).

Since (2.5.1) denotes a scalar problem, invariant imbedding appeared attractive, particularly, since only the final value $u(1)$ was desired. Because of the complicated expression for g no linearization was attempted. Since $y(1)$ is given and $u(1)$ to be found, the equation for $y(t)$ was chosen as base characteristic equation. Without further analysis it was assumed "on physical grounds" that the corresponding invariant imbedding equation

$$\partial u/\partial\xi - (\partial u/\partial y)g(u, y) = g(u, y)(\alpha y - \beta), \qquad u(0, y) = \hat{u} \tag{2.5.2}$$

has a solution over a sufficiently large domain, which was to be found numerically. For this example the conservation scheme was chosen. A triangular mesh with base at $\xi = 0$ extending from Y_1 to Y_2 and apex at $(1, \hat{y})$ was placed on the (ξ, y)-plane and the solution of (2.5.2) was advanced from one ξ-level to the next according to formula (2.4.8). The algorithm was easy to implement and, as indicated below, gave good results as verified by back integration of the system (2.5.1) subject to $u(1) = u(1, \hat{y})$, $y(1) = \hat{y}$. Table 2.5.1 lists the values obtained for a typical oil refining application. Here, $u(0)$ was found by integrating (2.5.1) subject to $u(1) = u(1, \hat{y})$, $y(1) = \hat{y}$ with a fourth-order Runge–Kutta method with a constant step size of 0.01.

Several observations apply to the system (2.5.1). In the reactor problem the parameter ε is positive so that the numerical scheme of Courant *et al.* is applicable provided u_y is replaced by a backward difference for $\xi \in [0, 1]$, $y \geq 0$, and provided that $\Delta\xi/\Delta y$ is sufficiently small. Second, we observe that the characteristic $\{u(\xi), y(\xi)\}$ through the point $(\hat{u}, 0)$ is a constant. Hence (2.5.1) can be solved subject to the boundary value problems $u(0, y) = \hat{u}$, $u(\xi, 0) = \hat{u}$. To this formulation, the unconditionally stable

Table 2.5.1

SOLUTION OF THE ARRHENIUS REACTION EQUATIONS[a]

N	$\hat{u}$	$\hat{y}$	$u(1, \hat{y})$	$u(0) - \hat{u}$
1	650	1.0	662.23	-0.57
10	650	1.0	663.07	0.10
20	650	1.0	663.02	0.06
30	650	1.0	662.99	0.04
40	650	1.0	662.98	0.03
50	650	1.0	662.97	0.02

[a] Constants: $\alpha = 2.13 \times 10^{-3}$; $\beta = 1.13 \times 10^{2}$; $\gamma = 2.34 \times 10^{4}$; $\varepsilon = 1.50 \times 10^{8}$; $Y_1 = 0$; $Y_2 = 2$; $\Delta \xi = \Delta y = 1/N$.

method of Shampine and Thompson (1970) applies. None of these alternative methods was evaluated (N.5.2). ∎

Example 2.5.2. Biological Reaction Problems

As a second example, let us consider a problem from the field of water purification treatment. Atkinson and Daoud (1968) have presented a model for the diffusion of a chemical reactant in a bed of microorganisms subject to a Michaelis–Menten reaction mechanism. This model requires the evaluation of the nonlinear two-point boundary value problem

$$v'' - \alpha v/(1 + \beta v) = 0, \qquad v'(0) = 0, \quad v(T) = \gamma$$

for a range of positive values for α, β, and γ. Only nonnegative concentrations ($v \geq 0$) are allowed. As in the chemical reactor problem, one boundary value rather than the complete solution is of interest. In this case the flux $v'(T)$ at the surface of the microorganism layer is desired.

It will be convenient to introduce the new variables

$$x = \beta v, \qquad t = \alpha^{1/2} \xi$$

The boundary value problem can be written in terms of x and t as

$$x'' - x/(1 + x) = 0, \qquad ' = d/dt$$
$$x'(0) = 0, \qquad x(\hat{T}) = \hat{x}, \qquad \hat{T} = T\sqrt{\alpha}, \qquad \hat{x} = \beta\gamma$$

When converted into a system of first-order equations, this problem shows a number of desirable features. We observe from

$$u' = x/(1 + x), \qquad u(0) = 0$$
$$x' = u, \qquad\qquad x(\hat{T}) = \hat{x}$$
$$(2.5.3)$$

that the characteristics $\{u(t), x(t)\}$ exist over $[0, 1]$ for arbitrary nonnegative initial values. Furthermore, we see that in the terminology of Section 2.3, we have

$$\partial F/\partial x = 1/(1 + x)^2, \qquad \partial G/\partial u = 1$$

Hence it follows from Theorem 2.3.4 that the Cauchy problem

$$\partial u/\partial t + (\partial u/\partial x)u = x/(1 + x), \qquad u(0, x) = 0 \qquad (2.5.4)$$

has a global solution for t, $x \geq 0$. It is also readily verified that the characteristic through $(0, 0)$ has the solution $u(t) = v(t) \equiv 0$ so that the solution $u(t, x)$ of (2.5.4) satisfies the boundary conditions

$$u(0, x) = u(t, 0) = 0$$

As a consequence, the unconditionally stable method of Shampine and Thompson is applicable for the evaluation of

$$u(\hat{T}, \hat{x}) = x'(\hat{T}) = \frac{\beta}{\sqrt{\alpha}} (dv/d\xi)(T)$$

Figures 2.5.1 and 2.5.2 show the results of a sample calculation for the special case $\alpha = T = \gamma = 1$. A mesh spacing of $\Delta t = 10^{-2}$ and $\Delta x = 10^{-2}$ was found to be adequate. Figure 2.5.1 shows the solution $u(1, \beta\gamma) = \beta \, dv/d\xi$ and the flux $dv/d\xi = (1/\beta)u(1, \beta\gamma)$ as a function of β. They were obtained by linear interpolation between the computed values of $u(1, x_j)$. The plotted value for $dv/d\xi$, $\xi = 1$, $\beta = 0$ was computed as $dv/d\xi = (e^2 - 1)/(e^2 + 1)$ from the closed form solution of (2.5.3). The same value is obtained by taking the right sided limit of $dv/d\xi$ at $\beta = 0$ in Fig. 2.5.1. Figure 2.5.2 shows the curve $u(0, x)$ obtained by interpolating linearly between the values $u_j(0)$ which were found by back integrating (2.5.3) subject to $u(1) = u(1, x_j)$, $x(1) = x_j$ with a fourth-order Runge–Kutta method (Step 3′ of the invariant imbedding algorithm). The exact solution would be identically zero. For this problem the maximum error is approximately 3.2×10^{-4}. Note that a cut through the integral surface $u(t, x)$ of (2.5.4) along a line $x = $ constant provides a solution of the Michaelis–Menten equation for constant β and variable α.

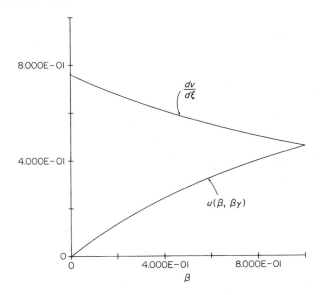

Fig. 2.5.1. Integral surface $u(1, \beta\gamma)$ and flux $dv/d\xi = (1/\beta)u(1, \beta\gamma)$ as a function of β obtained from (2.5.4) with the semi-implicit method (2.4.7).

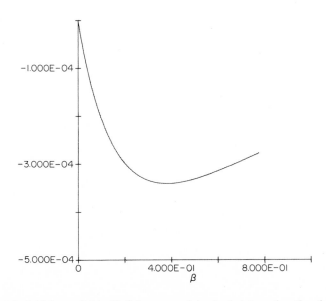

Fig. 2.5.2. Initial manifold $u(0, \beta\gamma)$ computed by back integrating the characteristic equations (2.5.3) with a Runge–Kutta method subject to the initial values $(u(1, \beta\gamma), \beta)$.

It may be of interest to contrast the invariant imbedding approach to the method of finite differences for the Michaelis–Menten reaction equation. Keller (1968) uses a spherical rather than the linear model, but the same comments apply to both geometries. The nonlinearity of the problem coupled with finite difference approximations for the derivatives will require the solution of a nonlinear system of equations. If an iterative technique like Newton's method is used to solve the equations, care must be exercised to avoid convergence to nonphysical negative solutions $x(\xi)$ of (2.5.3). Because the integral surface $u(t, x)$ of (2.5.4) necessarily passes through the characteristic manifold $u(t) = 0$ over the base characteristics $x(t) = 0$, no base characteristic $x(t)$ subject to a nonnegative initial value can cross into the half plane $x < 0$. Hence, invariant imbedding will not yield negative solutions for physical boundary values. ∎

Example 2.5.3. The Thomas–Fermi Equation—A Caveat

The Thomas–Fermi method of atomic theory requires the solution of the following well-studied two-point boundary value problem

$$u'' = u(u/\xi)^{1/2}, \qquad u(0) = 1, \qquad u'(T) = u(T)/T \qquad (2.5.5)$$

The standard invariant imbedding approach applied to the equivalent first-order system

$$u' = x, \qquad\qquad u(0) = 1$$
$$x' = u(u/\xi)^{1/2}, \qquad u(T) = Tx(T)$$

requires solving

$$\partial u/\partial \xi + (\partial u/\partial x)u(u/\xi)^{1/2} = x, \qquad u(0, x) = 1 \qquad (2.5.6)$$

and determining a solution $\hat{x}$ of the equation

$$g(x) \equiv u(T, x) - Tx = 0 \qquad (2.5.7)$$

Equation (2.5.6) was integrated numerically over the domain $0 \le \xi \le 3$, $0 \le x \le 3$ according to the following hybrid formula

$$\frac{u_{n+1,i} - u_{n,i}}{\Delta\xi} + \frac{u_{n+1,i} - u_{n+1,i-1}}{\Delta x}\left[u_{n+1,i}\left(\frac{u_{n,i}}{\xi_{n+1}}\right)^{1/2}\right] = x_i, \qquad i = 1,\ldots, 3/\Delta x$$

$$\frac{u_{n+1,0} - u_{n,0}}{\Delta\xi} + \frac{u_{n,1} - u_{n,0}}{\Delta x}\left(\frac{u_{n,0}^3}{\xi_{n+1}}\right)^{1/2} = 0$$

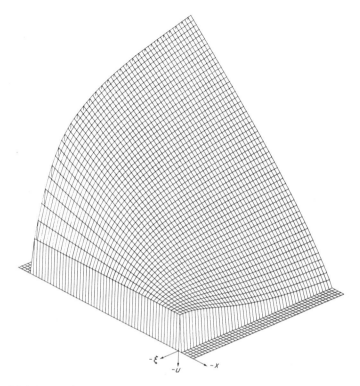

Fig. 2.5.3. Integral surface for the invariant imbedding equation (2.5.6) corresponding to the Thomas–Fermi equation over the square $0 \leq \xi \leq 3$, $0 \leq x \leq 3$.

The integral surface was found over this strip without difficulty. It is shown in Fig. 2.5.3 for an evenly spaced mesh of 50 points along each axis. A finer mesh did not lead to readily observable differences. Indeed, Fig. 2.5.4 shows the function g obtained through linear interpolation between the mesh points x_i for two different cases. Curve No. 1 is the result for a mesh size of $\Delta\xi = \Delta x = 3 \times 10^{-2}$, curve No. 2 that for a mesh size of $\Delta\xi = \Delta x = 3 \times 10^{-3}$. There is little apparent change in the computed results so that the finite difference solution of (2.5.6) would generally be considered acceptable. From Fig. 2.5.4 we obtain the solution

$$\hat{x} = 0.183$$

for equation (2.5.7).

Figure 2.5.5 shows the solution curve $u(\xi)$ of (2.5.5) obtained by integrating (2.5.5) subject to

$$u(3) = 3\hat{x}, \qquad x(3) = \hat{x}$$

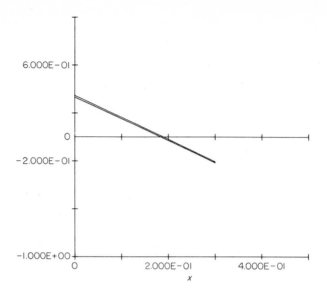

Fig. 2.5.4. Plot of the function g defined by (2.5.7). The upper curve was obtained when computing u with a mesh size of $\Delta\xi = \Delta x = 3 \times 10^{-2}$. The lower curve corresponds to $\Delta\xi = \Delta x = 3 \times 10^{-3}$.

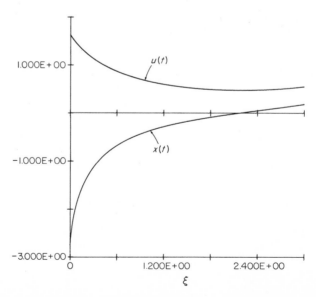

Fig. 2.5.5. Back integration of the Thomas–Fermi equation with a Runge–Kutta method.

with a fourth-order Runge–Kutta method. It is apparent that the computed solution is not acceptable. The surface characteristic $u(t)$ is supposed to pass through the initial manifold $u(0, x) = 1$ while the computed value is $u(0) = 1.62$. The reason for this discrepancy is readily discernable. We have integrated the invariant imbedding equation (2.5.6) numerically over the strip $D = \{(\xi, x): 0 \leq \xi \leq 3, 0 \leq x \leq 3\}$ and determined a solution $\hat{x} \in D$ from the boundary condition (2.5.7). However, back integration of (2.5.5) subject to $u(3) = u(3, \hat{x})$, $x(3) = \hat{x}$ shows that the base characteristic $x(t)$ lies outside this strip for $t \in [0, 2.17]$. Hence we have violated one of the fundamental requirements for the numerical integration of hyperbolic equations, namely: The region of integration must contain the domain of dependence for the point $(T, u(T, \hat{x}), \hat{x})$. In our first-order case this domain of dependence is the characteristic $\{u(t), x(t)\}$ through $(u(T, \hat{x}), \hat{x})$, and it leaves the surface u over D already at $t = 2.17$.

The numerical solution of the Thomas–Fermi equation with invariant imbedding was abandoned at this point because extending the strip D into the half plane $x < 0$ invariably led to negative values for some $u(\xi_i, x_j)$ which caused our algorithm to become invalid due to negative radicands. We concluded that invariant imbedding is not a good method to use for this problem (see N.5.3). ∎

2.6. LINEAR PROBLEMS

If we apply the method of invariant imbedding to the linear boundary value problem

$$u' = A(t)u + B(t)x + F(t), \qquad u(0) = fx(0) + a$$
$$x' = C(t)u + D(t)x + G(t), \qquad g_1u(T) + g_2x(t) = b \qquad (2.6.1)$$

where $u = (u_1, \ldots, u_m)^T$, $x = (x_1, \ldots, x_n)^T$ and where all given functions are continuous in t, then according to our algorithm, the missing boundary value $x(T)$ is found from the equation

$$g_1u(T, x) + g_2x = b$$

where $u(t, x)$ is the integral surface of the corresponding invariant imbedding equation

$$u_t(t, x) + u_x(t, x)[C(t)u + D(t)x + G(t)] = A(t)u + B(t)x + F(t)$$
$$u(0, x) = fx + a$$

In the discussion of partial differential equations with the same principal part (Section 1.2) it was established that for linear characteristic equations this integral surface can be written as an affine transformation

$$u(t, x) = U(t)x + v(t)$$

where U and v satisfy the Riccati equation (1.2.5) and the linear equation (1.2.6) respectively. Because of this simplification the algorithm of Section 2.4 can be restated for linear problems as follows:

Algorithm. The Method of Invariant Imbedding for Linear Boundary Value Problems

The boundary value problem

$$
\begin{aligned}
u' &= A(t)u + B(t)x + F(t), & u(0) &= fx(0) + a \\
x' &= C(t)u + D(t)x + G(t), & g_1u(T) + g_2x(T) &= b
\end{aligned}
\qquad (2.6.1)
$$

can be solved in the following four steps:

Step 1. Integrate the Riccati equation

$$U' = B(t) + A(t)U - UD(t) - UC(t)U, \qquad U(0) = f \quad (2.6.2)$$

over $[0, T]$.

Step 2. Integrate the linear equation

$$v' = [A(t) - U(t)C(t)]v - U(t)G(t) + F(t), \qquad v(0) = a \quad (2.6.3)$$

over $[0, T]$.

Step 3. Find the solution $\hat{x}$ of the equation

$$[g_1U(T) + g_2]x = b - g_1v(T) \qquad (2.6.4)$$

Step 4. Integrate the linear equation

$$x' = [C(t)U(t) + D(t)]x + C(t)v(t) + G(t), \qquad x(T) = \hat{x} \quad (2.6.5)$$

backward over $[0, T]$.

The functions $\{u(t) \equiv U(t)x(t) + v(t), x(t)\}$ are a solution of the boundary value problem (2.6.1).

Again, it may be impractical or impossible to store $U(t)$ and $v(t)$ to integrate (2.6.5). In this case Step 4 can be replaced by

Step 4'. Integrate the initial value problem

$$
\begin{aligned}
u' &= A(t)u + B(t)x + F(t), & u(T) &= U(T)\hat{x} + v(T) \\
x' &= C(t)u + D(t)x + G(t), & x(T) &= \hat{x}
\end{aligned}
\tag{2.6.6}
$$

It is straightforward to verify by differentiation and use of the defining equations for U and v that $u(t)$ is in fact a solution of the system (2.6.1). This verification is mechanical and does not rely on any concepts of characteristic theory. It will prove useful when applying the method of invariant imbedding to abstract boundary value problems (Chapter 5). It does, however, conceal the natural interpretation of $u(t, x)$ as an integral surface which allows the interpretation of invariant imbedding as an implicit shooting method.

Finally, it may be noted that in the terminology of Chapter 1 the functions $u(t)$ and $x(t)$ correspond to the surface and base characteristics, respectively. As was pointed out in Section 2.4 the linear boundary value problem may be given in the form

$$
y' = A(t)y + F(t)
$$

so that the identification of components of y with u and x is up to the user. The earlier comments apply to linear systems as well, and a judicious choice of u and x may materially simplify the computation (see Example 2.8.2). Some related remarks concern the reduction of higher-order linear equations to first-order systems. Specifically, consider the important class of Sturm–Liouville-type boundary value problems

$$
\begin{aligned}
(p(t)u')' - q(t)y &= F(t) \\
\alpha_1 u'(0) - \beta_1 u(0) &= \gamma_1 \\
\alpha_2 u'(T) - \beta_2 u(T) &= \gamma_2
\end{aligned}
$$

where $p(t)$ is a differentiable positive function on $[0, T]$. The simplest reduction would appear to be

$$
\begin{aligned}
u' &= x \\
x' &= \frac{q(t)}{p(t)} u - \frac{p'(t)}{p(t)} + \frac{F(t)}{p(t)}
\end{aligned}
$$

with the appropriate boundary conditions. Invariant imbedding is, of course, applicable to this formulation. If, however, we define a function z by the relation

$$u' = z/p(t)$$

then we obtain the equations

$$u' = z/p(t), \qquad \alpha_1 z(0)/p(0) = \beta_1 u(0) + \gamma_1$$
$$z' = q(t)u + F(t), \qquad \alpha_2 z(T)/p(T) = -\beta_2 u(T) + \gamma_2$$

It is apparent that the resulting invariant imbedding equations assume a somewhat simpler form. In particular, the linear term is missing in the Riccati equation. We remark in passing, that if z is taken to be the surface characteristic and u the base characteristic, we obtain the Riccati transformation

$$z(t) = R(t)u(t) + w(t)$$

and the invariant imbedding equations

$$R' = q(t) - (1/p(t))R^2, \qquad R(0) = \beta_1 p(0)/\alpha_1$$
$$w' = -(R(t)/p(t))w + F(t), \qquad w(0) = p(0)\gamma_1/\alpha_1$$

The boundary condition at $t = T$ and the Riccati transformation furnish the boundary value $u(T)$ as the solution of the equation

$$u(T) = [\alpha_2 R(T)/p(T) + \beta_2]^{-1}[\gamma_2 - \alpha_2 w(T)/p(T)]$$

The complete solution $u(t)$ is found from the reverse sweep

$$u' = (R(t)/p(t))u(t) + w(t)/p(t)$$

These imbedding equations are identical to the initial value formulation of Babuska *et al.* (1966) found by factoring the Sturm–Liouville operator. The existence of solutions for this problem is governed by Theorem 2.3.5.

2.7. NUMERICAL CONSIDERATIONS FOR LINEAR PROBLEMS

There is considerable competition to invariant imbedding when solving linear problems. Finite difference, Galerkin, and shooting methods are easily set up and come equipped with rather precise error bounds. The same is true for our initial value approach, and we shall present examples where invariant imbedding is actually superior, either in computing speed or mathematical stability, to some of the alternate methods. Of course, if the boundary value problem is to be solved on an analog computer then conversion to initial value problems is mandatory. In this case, stability considerations will dictate whether invariant imbedding or any of the other methods mentioned in Section 2.1 are to be chosen (N.7.1).

For the time being it will be assumed that the ordinary differential equations (2.6.2), (2.6.3), (2.6.5), or (2.6.6) are to be integrated with any of the many standard initial value techniques such as Euler's method, the trapezoidal rule (used extensively below), a Runge–Kutta method, etc. Some specific comments concerning the numerical integration of the imbedding equations may be found at the end of this section.

Before turning to the virtues of invariant imbedding let us describe some cases where the method is likely to run into difficulties. First and foremost is the problem of critical lengths which is reflected in unbounded solutions for the Riccati equation (2.6.2). They are usually due to the quadratic term of the equation rather than some isolated singularity of the coefficients. Difference equations will mirror the quadratic growth of the right-hand side of (2.6.2) so that the numerical solutions will blow up as well. While this behavior has the advantage that one cannot inadvertently integrate beyond the critical length, in which case invariant imbedding is not valid (in other words, invariant imbedding will not work in practice when it does not apply in theory), it causes difficulties if T is very near the critical length. For square systems this problem can generally be overcome by solving for $U^{-1}(t)$ instead of $U(t)$ provided, of course, that $U^{-1}(t)$ is well behaved near the singular point. This question is taken up in the next chapter in connection with interface and eigenvalue problems. If U is not a square matrix, invariant imbedding will fail for T near or beyond the critical length.

But even when the imbedding equations are well behaved, the method may not be practical. Integration of the $(m \times n)$-dimensional Riccati equation with a standard high order numerical routine such as the Runge–Kutta method may prove extremely time consuming for large matrices. Furthermore, storage of $U(t)$ over $[0, T]$ as required for the integration of (2.6.5) may be impossible because of memory size limitations. However, in contrast to the nonlinear case, the dimension of the base characteristic is of no importance for linear problems. The number of equations is largest when base and surface characteristics have the same dimension and smallest when one of these characteristics is one dimensional. The (backward) integration of Eq. (2.6.5) is further complicated by the fact that the Runge–Kutta routine will provide U and v only at the mesh points, while the integration of (2.6.5) with the same integrator will need intermediate values. This in turn will require costly interpolation.

Instead of the reverse sweep (2.6.5) one can integrate the initial value problem (2.6.6) for which $U(t)$ is not needed. However, avoidance of a pure initial value problem for (2.6.6) may well be the reason for using invariant imbedding in the first place. This brings us to one advantage of invariant imbedding. To be specific, let us consider in more detail the type of problem that is difficult to solve by integrating an initial value problem for (2.6.1) but which is readily handled with the imbedding technique. We shall use the example presented by Allen and Wing (1970). The equation

$$u'' - u = 0, \qquad u(0) = 1, \qquad u'(T) = -e^{-T} \qquad (2.7.1)$$

has the general solution $u(t) = C_1 e^t + C_2 e^{-t}$. The boundary conditions require that $C_1 = 0$. However, due to round-off errors, a numerical solution of (2.7.1) obtained by integrating the equations subject to $u(0) = 1$, $u'(0) = -1$ cannot entirely suppress the positive exponential term which will dominate the solution at $t = T$ for large T and make the computed result useless. On the other hand, application of invariant imbedding leads to the Riccati equation (1.2.7) with solution $U(t) = \tanh t$, while the solutions of (2.6.3) and (2.6.5) are exponentially decreasing. The positive exponential solution is completely eliminated in the invariant imbedding approach.

The stability of the forward–backward integration required by invariant imbedding has been discussed in detail by Babuska *et al.* (1966) in connection with scalar second- and fourth-order boundary value problems which were converted to initial value problems by factoring the differential operators. As indicated in Section 2.6 their equations are identical to (2.6.2)–(2.6.5).

These stability considerations also extend to systems of the following form

$$u' = A(t)u + B(t)x, \qquad x' = C(t)u - A^*(t)x \qquad (2.7.2)$$

where $-(A(t) + A^*(t))$, $B(t)$, and $C(t)$ are positive definite. It follows from

$$P(t)\begin{pmatrix} x \\ y \end{pmatrix} \equiv \begin{pmatrix} A(t) & B(t) \\ C(t) & -A^*(t) \end{pmatrix} \begin{pmatrix} x \\ y \end{pmatrix} = \lambda \begin{pmatrix} x \\ y \end{pmatrix}$$

that if λ is an eigenvalue and $(x, y)^{\mathrm{T}}$ the corresponding eigenvector, then $-\lambda$ and $(-y, x)^{\mathrm{T}}$ are the eigenvalue and eigenvector of P^*. Hence the spectrum of

$$\begin{pmatrix} A(t) & B(t) \\ C(t) & -A^*(t) \end{pmatrix}$$

has points in the positive half plane. An initial value problem for (2.7.2) is not stable in the Liapunov sense and the shooting method will have to contend with exponentially growing solutions. On the other hand, it is straightforward to verify that the invariant imbedding equations correspond-ing to (2.6.8) are Liapunov stable and hence easy to integrate.

The stability of the invariant imbedding equations is one—for an analog computer solution essential—benefit of invariant imbedding for problems of the type (2.6.1). The second important advantage is due to the fact that the Riccati equation (2.6.2) does not depend on the source functions F and G or on any boundary values except f. This property can lead to considerable economy when solving (2.6.1) for a variety of source terms and boundary inputs because the nonlinear Riccati equation has to be solved only once. The examples of the next section illustrate this point.

The numerical integration of the imbedding equations (2.6.2)–(2.6.6) is, at least for smooth right-hand sides, a standard application for the many integration schemes studied in the literature. These methods generally come equipped with rather precise error bounds and it is not difficult to relate such bounds to error estimates for the boundary value problem (2.6.1). As an illustration, let us derive an *a posteriori* error bound for a linear boundary value problem with simple boundary conditions. First, we shall state a comparison result for the solution of differential equations subject to different boundary values. Throughout this discussion we shall assume that the hypotheses of Theorem 1.1.3 apply to the coefficients so that the Riccati equation (2.6.2) has a bounded solution on $[0, T]$.

LEMMA 2.7.1. Let $\{u_i(t), x_i(t)\}$, $i = 1, 2$, be the solutions of (2.6.1) sub-ject to the boundary values $u_i(0) = fx_i(0) + a_i$, $g_1 u_i(T) + g_2 x_i(T) = b_i$. Then there exists a (generic) constant K such that

$$\| a_1(t) - u_2(t) \| \leq K(\| a_1 - a_2 \| + \| b_1 - b_2 \|)$$

$$\| x_1(t) - x_2(t) \| \leq K(\| a_1 - a_2 \| + \| b_1 - b_2 \|)$$

uniformly in t on $[0, T]$ provided $[g_2 + g_1 U(T)]$ has a bounded inverse.

Proof. The functions $u_3(t) = u_2(t) - u_1(t)$, $x_3(t) = x_2(t) - x_1(t)$ satisfy the boundary value problem

$$u_3' = A(t)u_3 + B(t)x_3, \qquad u_3(0) = fx_3(0) + (a_2 - a_1)$$

$$u_3' = C(t)u_3 + D(t)x_3, \qquad g_2 x_3(T) = g_1 u_3(T) + (b_2 - b_1)$$

The solution is given by the invariant imbedding equations as

$$u_3(t) = U(t)x_3(t) + v(t)$$

where $v(t) = \Phi(t, 0)(a_2 - a_1)$ and

$$x_3(t) = \psi(t, T)[g_2 + g_1 U(T)]^{-1}[-g_1 \Phi(T, 0)(a_2 - a_1) + b_2 - b_1].$$

Here Φ and ψ are the fundamental matrices of the linear systems

$$z' = [A(t) - U(t)C(t)]z, \qquad z' = [C(t)U(t) + D(t)]z$$

Since Φ and ψ are bounded, it follows that there exists some constant K such that

$$\| x_3(t) \| \leq K(\| a_2 - a_1 \| + \| b_2 - b_1 \|)$$
$$\| u_3(t) \| \leq K(\| a_2 - a_1 \| + \| b_2 - b_1 \|) \quad \blacksquare$$

Let us suppose that the invariant imbedding equations for U and v have been integrated numerically, however badly, over $[0, T]$ and that the boundary value $\hat{x}$ has been determined (possibly even guessed). The complete solution of the boundary value problem will now be found by back integration of the initial value problem (2.6.6.). We shall denote this numerical solution by $\{u_h(t), x_h(t)\}$ while the exact solution is $\{u_1(t), x_1(t)\}$. Then we can compare the true and computed solution.

THEOREM 2.7.1. Suppose that Eqs. (2.6.6) are integrated backward over $[0, T]$ with a numerical method of order h^p, $p > 0$, where h is the step length. Let the vector ε be defined by $\varepsilon = u_h(0) - fx_h(0) - a$. Then

$$\max\{\| u_h(t) - u_1(t) \|, \| x_h(t) - x_1(t) \|\} \leq K(h^p + \| \varepsilon \|)$$

for some constant K uniformly in t at the mesh points in $[0, T]$.

Proof. Let $u_2(t), x_2(t)$ be the exact solution of

$$u' = A(t)u + B(t)x + F(t), \qquad g_1 u_h(T) + g_2 x_h(T) = b$$
$$x' = C(t)u + D(t)x + G(t), \qquad x_h(T) \quad \text{as computed from} \quad (2.6.4)$$

Then

$$u_2(0) - fx_2(0) - a = u_h(0) + (u_2(0) - u_h(0)) - fx_h(0) + f(x_h(0) - x_2(0)) - a$$
$$= \varepsilon + [u_2(0) - u_h(0)] + f[x_h(0) - x_2(0)]$$

Since $u_h(t)$ and $x_h(t)$ were found with a method of order h^p we have

$$\| u_2(0) - u_h(0) \| \leq Kh^p \qquad \text{and} \qquad \| x_h(0) - x_2(0) \| \leq Kh^p$$

Applying Lemma 2.7.1 to $\{u_i(t), x_i(t)\}$, $i = 1, 2$, and noting that $b_1 = b_2 = b$ we see that

$$\max\{\| u_1(t) - u_2(t) \|, \| x_1(t) - x_2(t) \|\} \leq K(\| \varepsilon \| + (1 + \| f \|)h^p)$$

for some constant K. The conclusion of the theorem follows from the triangle inequality applied to $\| u_h - u_1 \| = \| (u_h - u_2) - (u_1 - u_2) \|$. ∎

It would appear possible to derive a comparison theorem similar to Lemma 2.7.1 for nonlinear boundary value problems which are "close" to linear systems. At any rate, whenever such a result is available, Theorem 2.7.1 can be applied to obtain an *a posteriori* bound. This bound is influenced only by the choice of integrator for (2.2.6) and not by the initial values $(u(T, \hat{x}), \hat{x})$; hence it would not reflect the low order of convergence available for the finite difference methods of section 2.4. For the linear problems the imbedding equations (2.6.2) and (2.6.3) can also be integrated with a numerical method of order h^p. In this case the error ε is itself of order h^p as the next theorem shows.

THEOREM 2.7.2. Suppose Eqs. (2.6.2), (2.6.3), and (2.6.6) are integrated with a method of order h^p, $p > 0$, then

$$\| u(t) - u_h(t) \| \leq Kh^p, \qquad \| x(t) - x_h(t) \| \leq Kh^p$$

for sufficiently small h and some (generic) constant K.

Proof. We know from the hypotheses that $\| U_h(T) - U(T) \| \leq Kh^p$ and $\| v_h(T) - v(T) \| \leq Kh^p$ for sufficiently large K. These estimates are used to bound the difference $\| \hat{x} - \hat{x}_h \|$ as computed from (2.6.4); we see that

$$\hat{x} = [g_1 U(T) + g_2]^{-1}[b - g_1 v(T)]$$

and

$$\begin{aligned} \hat{x}_h &= [g_1 U_h(T) + g_2]^{-1}[b - g_1 v_h(T)] \\ &= [I + (g_1 U(T) + g_2)^{-1} g_1(U_h(T) - U(T))]^{-1}[g_1 U(T) + g_2]^{-1} \\ &\quad \times [b - g_1 v(T) - g_1(v_h(T) - v(T))] \end{aligned}$$

It follows that $\hat{x}_h$ can be expressed as

$$\hat{x}_h = [I + P]^{-1}[g_1 U(T) + g_2]^{-1}[b - g_1 v(T) - \varepsilon]$$

where $\| P \| \leq Kh^p$ and $\| \varepsilon \| \leq Kh^p$ for some constant K. This expression leads to

$$\hat{x} - \hat{x}_h = [I - (I + P)^{-1}][g_1 U(T) + g_2]^{-1}[b - g_1 v(T)]$$
$$+ [I + P]^{-1}[g_1 U(T) + g_2]^{-1}\varepsilon$$

We shall assume that h is small so that $\| P \| < 1$. By Banach's lemma

$$\| I - (I + P)^{-1} \| = \| P(I + P)^{-1} \| \leq \| P \|/(1 - \| P \|).$$

Since we have assumed throughout that $[g_1 U(T) + g_2]^{-1}$ is bounded we find

$$\| \hat{x} - \hat{x}_h \| \leq \| P \| \frac{1}{1 - \| P \|} K + \frac{K}{1 - \| P \|} \| \varepsilon \| \leq Kh^p$$

Next, it may be observed that the exact solution $\{u_2(t), x_2(t)\}$ of (2.6.6) subject to $(U_h(T)\hat{x}_h + v_h, \hat{x}_h)$ differs from the correct solution $\{u(t), x(t)\}$ of (2.6.1) only by Kh^p because both satisfy the same differential equations. Hence $\| u_2(t) - u(t) \| \leq Kh^p$ and $\| x_2(t) - x(t) \| \leq Kh^p$. Since for the computed solution $\{u_h(t), x_h(t)\}$ the inequalities

$$\| u_2(t) - u_h(t) \| \leq Kh^p, \qquad \| x_2(t) - x_h(t) \| \leq Kh^p$$

are valid by hypothesis, the conclusion of the theorem follows again from the triangle inequality. ∎

It may be noted that if the reverse sweep (Step 4) is used, the solution $u_h(t) = U_h(t)x_h(t) + v_h(t)$ will always satisfy the boundary condition at $t = 0$ regardless of how incorrectly the invariant imbedding equations have been integrated. In this case an *a posteriori* error bound is not available. On the other hand, if the equations are integrated correctly, it is possible to derive an *a priori* error estimate again by determining first the difference between the true and computed boundary values at $t = T$ and then comparing the true and computed base characteristic $x(t)$. At first sight, the sweep approach might be expected to yield superior accuracy because at least one boundary condition is satisfied. However, as will become apparent, for higher order methods we are faced with interpolation problems. We shall suppose that the boundary values for (2.6.1) are given as

$$u(0) = a, \qquad x(T) = b$$

The following theorem shows the improvement possible with the sweep method.

THEOREM 2.7.2. Let the invariant imbedding equations (2.6.2, 3, and 5) have bounded solutions and suppose that these equations are integrated numerically with Euler's method. Then there exists a (generic) constant K such that

$$\| u(t) - u_h(t) \| \leq Kh\phi_1(t), \qquad \| x(t) - x_h(t) \| \leq Kh\phi_2(T - t)$$

where ϕ_i, $i = 1, 2$ are nonnegative continuous functions on $[0, T]$ such that $\phi_i(0) = 0$.

Proof. The exact and computed solutions of Eqs. (2.6.2) and (2.6.3) satisfy

$$U_h(t) - U(t) = \varepsilon(t), \qquad v_h(t) - v(t) = \delta(t)$$

at the mesh points where $\| \varepsilon(t) \| \leq KhE_K(t)$, $\| \delta(t) \| \leq KhE_K(t)$ for a sufficiently large constant K and where E_K is the Lipschitz function defined by $E_K(t) = (e^{Kt} - 1)/K$. If we define ε and δ as continuous functions over $[0, T]$ by linear interpolation and call x_2 the exact solution of

$$x' = [C(t)U(t) + D(t)]x + C(t)\varepsilon(t)x + C(t)v(t) + C(t)\delta(t)$$
$$x(T) = b$$

and if we set $\gamma(t) = x_2(t) - x(t)$, then it follows from the boundedness of $x(t)$ that

$$\| \gamma(t) \| \leq Kh(T - t)$$

for sufficiently large K. The Euler approximation x_h to x_2 satisfies

$$\| x_h(t) - x_2(t) \| \leq KhE_K(T - t)$$

Continuing these estimates, we obtain

$$\| x_h(t) - x(t) \| \leq \| x_h(t) - x_2(t) \| + \| x_2(t) - x(t) \| \leq Kh\phi_2(t)$$
$$\| u_h(t) - u(t) \| = \| U_h(t)x_h(t) + v_h(t) - U(t)x(t) - v(t) \|$$
$$\leq \| U(t) + \varepsilon(t) \| \| x_h - x \| + \| \varepsilon(t) \| \| x(t) \|$$
$$\leq Kh\phi_1(t) \qquad \text{for sufficiently large } K \quad \blacksquare$$

Exactly the same development will go through for the trapezoidal rule because no values of $U(t)$ and $v(t)$ are needed between mesh points. If interpolated values for $U(t)$ and $v(t)$ are needed for the integration of (2.6.5), then the order of the integrator is retained by the solution of the boundary value problem only if the interpolation is carried out to the same order.

Many of the integrations throughout this monograph were carried through with a fourth-order Runge–Kutta method, and in general no difficulties were experienced. Occasionally, the formulation of the boundary value problem (2.6.1) already contained inherent errors (such as discretization errors when applying the method of lines (Example 2.8.2)) which negated the need for a highly accurate solution of (2.6.1). In this case the trapezoidal rule proved particularly efficient whenever the Riccati equation (2.6.2) could be solved readily.

To conclude this discussion of the numerical aspects, let us consider the integration of the Riccati equation (2.6.2). If the equation is one dimensional, the problem is readily solvable. Kamke (1943) lists a number of analytical solutions for various forms of the Riccati equation. The formulas presented in Table 2.7.1 have found repeated application in our work. There exists an amazing variety of closed form solutions for various Riccati equations and the reader is referred to the comprehensive compilation of Kamke. Of course, scalar problems are also readily integrated numerically. If the trapezoidal rule is of sufficient accuracy, a particularly fast algorithm

Table 2.7.1

SOME CLOSED FORM SOLUTIONS OF THE SCALAR RICCATI EQUATION[a]

Equation	Solution
$y' + ay^2 = b$ $y(\xi) = \eta$	$y(t) = \dfrac{\eta \sqrt{ab}\,(e^\gamma + e^{-\gamma}) + b(e^\gamma - e^{-\gamma})}{\sqrt{ab}\,(e^\gamma + e^{-\gamma}) + a\eta(e^\gamma - e^{-\gamma})}$ $\gamma = \sqrt{ab}\,(t - \xi), \qquad a, b \neq 0$ $y(t) = \dfrac{\eta}{1 + a\eta(t - \xi)}, \qquad b = 0$
$y' = (Ay - a)(By - b)$ $aB - bA \neq 0$	$y(t) = \dfrac{C_1 b \exp(aB - bA)t - C_2 a}{C_1 B \exp(aB - bA)t - C_2 A}$
$y' + ay^2 + \dfrac{2}{t}\,y + b = 0$	$y(t) = u(t) - \dfrac{1}{at}$ where $u(t)$ is the solution of $u' = -b - au^2$

[a] From Kamke (1943).

results because the nonlinear difference equation is quadratic and can be solved in closed form.

For the numerical solution of the matrix Riccati equation one generally has to resort to explicit numerical schemes. Since matrices of high order arise naturally, for example during the numerical solution of partial differential equations with the method of lines (see the following), such techniques will be costly in time and storage. As noted above, this feature is one of the limitations on invariant imbedding. Frequently, however, the Riccati equation has, or can be reduced to, the special form

$$U' = B - UCU, \qquad U(0) = f \qquad (2.7.3)$$

where B and C are constant $n \times n$ matrices. Let us suppose also that the following hypotheses can be met:

(i) B and C are nonsingular and BC has distinct eigenvalues;

(ii) the matrices BC and fC commute.

A nontrivial example where these conditions are satisfied is given in the next section.

We now propose to solve (2.7.3) as follows. First, the Riccati equation is post multiplied by the constant matrix C to yield

$$(UC)' = BC - (UC)^2, \qquad U(0)C = fC \qquad (2.7.4)$$

Define $V(t) = R^{-1}U(t)CR$, where R is the matrix whose columns are the eigenvectors of BC. Because of (ii), R will simultaneously diagonalize BC and fC. Post- and premultiplication of (2.7.4) by R and R^{-1} leads to the diagonal system

$$V' = \text{diag}\{\lambda_i\} - V^2, \qquad V(0) = \text{diag}\{\eta_i\} \qquad (2.7.5)$$

where $\{\lambda_i\}$ and $\{\eta_i\}$ are the eigenvalues of BC and fC.

Note that the n-dimensional Riccati equation (2.7.5) actually consists of n uncoupled scalar equations whose solutions are given in Table 2.7.1. Hence $V(t)$ is known and $U(t)$ is readily found from

$$U(t) = RV(t)C^{-1}R^{-1}$$

There exist many efficient computer routines for finding the eigenvectors and eigenvalues of arbitrary $n \times n$ matrices so that it generally is not difficult to compute the matrices R and $\text{diag}\{\lambda_i\}$, $\text{diag}\{\eta_i\}$.

2.8. SOLVING POTENTIAL AND DIFFUSION PROBLEMS

Invariant imbedding is a good method to use for potential problems to which the stability considerations of the preceding section apply. The following examples are meant to illustrate this point.

Example 2.8.1. A Standing Wave

As a first example, let us consider the linear problem

$$u'' = -u, \qquad u(0) = u(\pi) = 0 \tag{2.8.1}$$

Strictly speaking, this is not a diffusion equation. However, it will be treated as a special case of nonlinear Poisson equations considered by Courant and Hilbert (1962, II, p. 369). We are interested in finding the maximal solution u such that $|u| \leq N$. We note that straightforward invariant imbedding is not applicable because $[0, \pi/2]$ is the critical interval. However, as shown in the reference, this maximal solution can be found as the limit of a monotone sequence $\{u_m\}$, where u_m satisfies the boundary value problem

$$u_m'' - Nu_m = -(1 + N)u_{m-1}, \qquad u_m(0) = u_m(\pi) = 0 \tag{2.8.2}$$

with $u_0 = N(\pi - x)x$. Problem (2.8.2) when written in first-order form is amenable to solution by invariant imbedding. Since boundary values for u are prescribed, $u(t)$ will be chosen as surface characteristic. According to the recipe of Section 2.6, we determine u_m from the equations

$$U' = 1 - NU^2, \qquad\qquad\qquad\qquad U(0) = 0$$
$$v_m' = -NU(t)v_m + U(t)(1 + N)u_{m-1}, \qquad v_m(0) = 0$$
$$x_m(\pi) = -v_m(\pi)/U(\pi)$$
$$x_m' = NU(t)x_m + Nv(t) - (1 + N)u_{m-1}(t)$$
$$u_m(t) = U(t)x_m(t) + v_m(t)$$

The Riccati equation has the solution $U(t) = N^{-1/2} \tanh N^{1/2}t$, hence we only have to solve the last four equations for $m = 1, 2, \ldots$. A numerical solution with the trapezoidal rule yielded the iterates shown in Fig. 2.8.1 (N.8.1). ∎

Several observations should be brought out. The solution (2.8.2) by the method of particular integrals is similar to the Stefan problem discussed in Section 2.10 because initial value problems have to be integrated repeated-

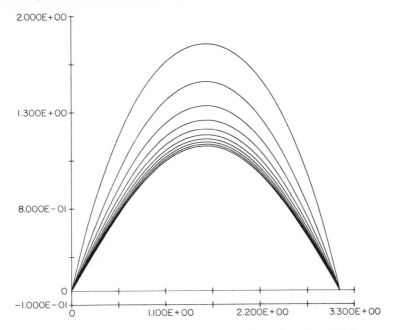

Fig. 2.8.1. Monotonely decreasing iteration for the (one-dimensional) Poisson equation (2.8.1).

ly which depend on the solution of the previous iteration. This approach may well be unstable because of the positive exponential fundamental solution. Second, while presented for a linear potential problem, the method is equally applicable to systems of the form

$$u'' = f(t, u)$$

where u is an m-dimensional vector and f is a bounded function. Such systems will arise if the Poisson equation

$$\Delta u = f(y, u), \qquad y = (y_1, y_2) \tag{2.8.3}$$

is discretized with the method of lines. Again, the Riccati equation will not depend on the iteration parameter m and has to be solved only once. The higher dimensional the system, the more saving in computer time is incurred by solving the Riccati equation only once and the more attractive invariant imbedding becomes. Of course, the modified successive substitution iteration for (2.8.3) analogous to iteration (2.8.2) may converge much more slowly than the Newton-type iteration

$$[\Delta - f_u(y, u_m)](u_{m+1} - u_m) = -\Delta u_m + f(y, u_m)$$

However, invariant imbedding at the $(m + 1)$th pass will now require the solution of a Riccati equation depending on u_m so that the solution of the linearized problem takes considerably more effort. Finally, it may be interesting to speculate to what extent an integration analogous to (2.8.2) can overcome critical length restrictions inherent in some invariant imbedding formulations. We remark in passing that an application of the iteration (2.8.2) with $f = u/(1 + u)$ and $N = 0$ produced a solution of the Michaelis–Menten equations (Example 2.5.2) in three iterations.

Example 2.8.2. Flow in a Fractured Oil Reservoir

When the flow of a slightly compressible fluid in a fractured oil reservoir is described mathematically, a rather unusual linear boundary value problem for the diffusion equation is obtained because second-order tangential derivatives occur on the boundary. The derivation of the applicable flow equation is described in detail by Cannon and Meyer (1971) so that only a short summary need be given here. Consider a circular reservoir with a well at its center and two narrow fractures extending from the well into the reservoir. Figure 2.8.2 shows the geometry under consideration. If the

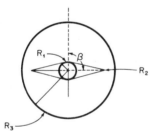

Fig. 2.8.2. Circular reservoir with two symmetric fractures.[†]

pressures on the well and reservoir boundary are held fixed, then the transient pressure behavior in the reservoir and fracture can be approximated by the variational equation (the integrated diffusion equation with a "narrow fracture" approximation)

$$\int_{\Omega} [k\,\nabla\xi \cdot \nabla u + \phi\mu c\xi(\partial u/\partial t) + \xi F(t, r)]r\,d\theta\,dr + \int_{\Gamma} [k(r)(\partial\xi/\partial r)(\partial u/\partial r)$$

$$+ \phi\mu c(\partial u/\partial t) + k\,\nabla\xi(\partial\psi/\partial r) + \xi F(t, r)]h(r)\,dr = 0 \qquad (2.8.4)$$

where ϕ is porosity of the formation, μ is viscosity of the fluid, c is compressibility of the fluid, k is permeability of the formation, $k(r)$ is permeability along the fracture, $h(r)$ is fracture width at distance r. The function ψ is defined as

$$\psi(t, r) = \frac{P_3(t) - P_1(t)}{R_3 - R_1} + P_1(t)$$

where $P_3(t)$ is pressure at drainage radius R_3, $P_1(t)$ is pressure at well radius R_1, and where $\xi(t, r)$ is a smooth test function with the proper boundary behavior. The function F is given by

$$F(t, r) = \phi\mu c \; \partial\psi/\partial t$$

If the segment Ω is transformed into a rectangle and the time derivatives are replaced by first-order backward differences, and if the solution is approximated by a Galerkin by lines technique exactly as described in Cannon and Meyer (1971), the following two-point boundary value problem results.

$$Au'' - Bu = F_n(y)$$
$$u'(Y) = 0, \qquad Y = \pi/2 \qquad (2.8.5)$$
$$Cu'(0) + Du(0) = g_n$$

Here A and B are positive definite constant $N \times N$ matrices, while C and D are semidefinite. Only the vectors F_n and g_n depend on the solution from the preceding time level. The vector u furnishes the pressure at the nth time level at N different radii. For a detailed discussion of Eq. (2.8.4), its properties and the reduction to (2.8.5), the reader is referred to the reference cited. Our aim here is to show how, after some transformations, invariant imbedding is well suited for the numerical solution of (2.8.5).

In order to bring (2.8.5) into standard form, we let $w = Au$ and $z' = w$ and use the equivalent first-order system.

$$z' = BA^{-1}w + F_n(y), \qquad\qquad z(Y) = 0$$
$$w' = z, \qquad\qquad CA^{-1}z(0) + DA^{-1}w(0) = g_n \qquad (2.8.6)$$

Invariant imbedding is immediately applicable to (2.8.6), but, as is readily verified, would lead to an $N \times N$ matrix Riccati equation. Since $N = 50$ is not unreasonable for practical computations, we would be faced with 2500 initial value problems. Fortunately, it turns out that the eigenvalues of BA^{-1} are distinct, hence BA^{-1} can be diagonalized by the corresponding matrix of eigenvectors. Let

$$R^{-1}BA^{-1}R = \text{diag}\{\lambda_i\}$$

and set

$$p(y) = R^{-1}z(y), \qquad q(y) = R^{-1}w(y)$$

Then (2.8.6) is equivalent to

$$
\begin{aligned}
p' &= \mathrm{diag}\{\lambda_i\}q + R^{-1}F_n(y), & p(Y) &= 0 \\
q' &= p, & CA^{-1}Rp(0) + DA^{-1}Rq(0) &= g_n
\end{aligned}
\tag{2.8.7}
$$

Invariant imbedding is to be applied to this system. The boundary condition $p(Y) = 0$ would suggest that $p(y)$ be considered the surface characteristic and that the integration proceed initially backward from $y = Y$ to $y = 0$. From the affine transformation

$$p = U(y)q + v(y) \tag{2.8.8}$$

we obtain the defining equations

$$U' = \mathrm{diag}\{\lambda_i\} - U^2, \qquad U(Y) = 0 \tag{2.8.9}$$

$$v' = U(y)v + R^{-1}F_n(y), \qquad v(Y) = 0 \tag{2.8.10}$$

The boundary value $q(0)$ is found from the linear system

$$[CA^{-1}RU(0) + DA^{-1}R]q(0) = g_n - CA^{-1}Rv \tag{2.8.11}$$

With $q(0)$ given the complete solution $u(y) = A^{-1}Rg(y)$ is found by integrating (2.8.8), i.e.,

$$p = q' = U(y)q + v(y)$$

subject to the computed initial value $q(0)$.

We observe that (2.8.9) is a diagonal matrix Riccati equation not depending on n, whose solution, for given λ_i, can be found from Table 2.7.1. Furthermore, if storage space permits, the matrix $[CA^{-1}RU(0) + DA^{-1}R]$ need be inverted only once since it also is independent of n. Note that if it is preferred to integrate the equations for U and v forward from 0 to Y, then (2.8.9) and (2.8.10) are subject to the initial values

$$U(0) = -R^{-1}AC^{-1}DA^{-1}R$$

$$v(0) = R^{-1}AC^{-1}g_n$$

The closed form solution (1.3.5) does not apply to this system because the diagonal matrices in (2.8.9) do not commute with the initial value $U(0)$.

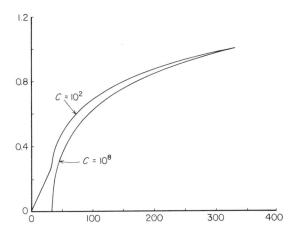

Fig. 2.8.3. Pressure distribution along the fracture for two representative reservoir calculations.

Here we have an application where the proper choice of the initial manifold substantially simplifies the computation.

Figure 2.8.3 shows the pressure distribution along a fracture in two steady-state calculations with representative physical parameters and $N = 40$. Figure 2.8.4 shows the corresponding normal pressure gradients which drive the oil into the reservoir. In this problem Chapeau functions were used for the partial Galerkin approximation (see Cannon and Meyer, 1971) while the initial value problems were integrated with the trapezoidal rule with step size Y/N. ▌

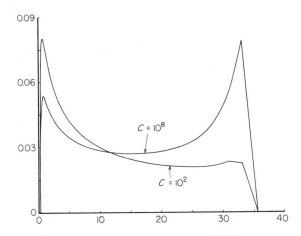

Fig. 2.8.4. Normal derivative $(1/r)(\partial p/\partial \theta)(r, 0)$.

2.9. NONLINEAR BOUNDARY CONDITIONS

Invariant imbedding was formulated in Section 2.2 for general nonlinear systems and nonlinear boundary values. For small systems where a numerical integration of the invariant imbedding equation is feasible, the numerical algorithm for generating the integral surface will generally be unaffected by the specific form of the initial manifold. Difficulties may arise when a non-linear second given boundary condition (i.e., Eq. (2.4.3)) is to be satisfied.

If the initial manifold and the characteristic equations are linear, this difficulty is no longer present. Consequently, invariant imbedding is well suited for the following class of problems.

$$
\begin{aligned}
u' &= A(t)u + B(t)x + F(t), & u(0) &= fx(0) + a \\
x' &= C(t)u + D(t)x + G(t), & g(u(T), x(T)) &= 0
\end{aligned}
\tag{2.9.1}
$$

where g is an arbitrary nonlinear continuously differentiable function. To elaborate, it follows from the discussion of linear equations in Section 2.6 that the boundary value $u(T)$ has the representation

$$
u(T) = U(T)x + v(T)
$$

where $U(t)$ and $v(t)$ are solutions of the initial value problems (2.6.2) and (2.6.3). Provided $U(t)$ has a solution over $[0, T]$ we can conclude that every distinct solution $\hat{x}$ of

$$
g(U(T)x + v(T), x) = 0
\tag{2.9.2}
$$

determines a distinct solution of (2.9.1), namely the characteristic through $(U(T)\hat{x} + v(T), \hat{x})$.

Here $U(T)$ and $v(T)$ are a given matrix and vector so that (2.9.2) is an analytic expression in the unknown x. If no closed form solution is obtainable, one may try approximation or iterative solution techniques to determine a root $\hat{x}$. Numerous theorems concerning the solvability of nonlinear systems are known (Ortega and Rheinboldt, 1970) which can guarantee the success of an iterative solution of (2.9.2). The structure and properties of g will determine which methods are to be tried.

Example 2.9.1. Stefan–Boltzmann Radiation

As an illustration of a linear system with a nonlinear boundary condition, let us consider the steady-state Stefan–Boltzmann radiation problem

$$
u'' - u = 0, \qquad u(0) = 1, \qquad u'(T) = -u^4
$$

The general recipe of Section 2.6, when applied to the first-order equivalent system, says that $u(T) = U(T)u'(T) + v(T)$, where U and v are the solutions of (2.6.2) and (2.6.3) which in this case assume the form

$$U' = 1 - U^2, \qquad U(0) = 0$$

and

$$v' = -U(t)v, \qquad v(0) = 1$$

The solution of these equations is seen to be $U(t) = \tanh t$ and $v(t) = \operatorname{sech} t$. The missing boundary value $u'(T)$ is a root of the equation

$$F(x) \equiv x + (x \tanh T + \operatorname{sech} T)^4 = 0$$

It is readily verified that $F(-\operatorname{csch} T) < 0$ and $F(0) > 0$ and that $F''(x) > 0$ on $[-\operatorname{csch} T, 0]$. Hence Newton's iteration

$$x_{n+1} = x_n - F(x_n)/F'(x_n), \qquad x_0 = 0 \tag{2.9.3}$$

will converge monotonically to a root $\hat{x}$ on $[-\operatorname{csch} T, 0]$. For $T = 1$, the root was computed to be

$$u'(1) = -0.1044548$$

and the complete solution is

$$u(t) = c_1 e^t + c_2 e^{-t} \tag{2.9.4}$$

where

$$c_1 + c_2 = 1, \qquad c_1 e - c_2 e^{-1} = u'(1) = -0.1044548$$

or

$$c_1 = 0.0854, \qquad c_2 = 0.9146$$

Figure 2.9.1 is a plot of this solution. We note that Eq. (2.9.3) has a second solution since $F(x) \to +\infty$ as $x \to -\infty$. For illustration, this solution is indicated as well. It was found from (2.9.3), starting with $x_0 = -4.0$, and from (2.9.4). ∎

In conclusion we see that invariant imbedding yields the same advantage of decoupling the boundary condition which is present when the radiation problem is solved by finite differences and the resulting algebraic system is solved by Gaussian elimination where only the last equation is nonlinear [see the discussion by Ciarlet *et al.* (1968)]. Analogous considerations will apply to higher order linear systems with one nonlinear boundary condition.

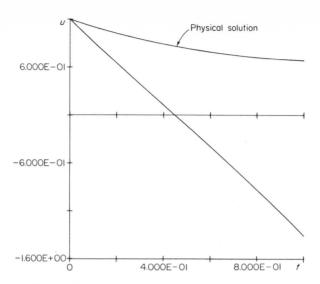

Fig. 2.9.1. Steady-state Stefan–Boltzmann radiation.

2.10. FREE BOUNDARY PROBLEMS

There exists an important class of problems where the location of the boundary (i.e., the length of the interval of integration) has to be determined as part of the solution for two-point boundary value problems. To this class belong such diverse topics as minimum time problems in the calculus of variations and time discretized one-phase Stefan problems. We shall assume first that the problem is formulated for linear inhomogeneous equations and a linear boundary condition at the fixed end. This will simplify the exposition and bring out the relation to the preceding section. Invariant imbedding is, of course, also applicable to nonlinear free boundary value problems. This topic is taken up at the end of this section.

We shall start with the system

$$u' = A(t)u + B(t)x + F(t), \qquad u(0) = fx(0) + a$$
$$x' = C(t)u + D(t)x + G(t) \tag{2.10.1}$$

with the free boundary $t = T$ to be determined such that

$$g(u(T), x(T)) = 0, \qquad h(u(T), x(T)) = 0 \tag{2.10.2}$$

where f is an $m \times n$ matrix, g is an n-dimensional vector, and h is a scalar function. Thus, there are $m + n + 1$ equations to determine the $m + n$

contants of integration $(u(T), x(T))$ and the location $t = T$ of the free boundary.

The treatment of problem (2.10.1) and (2.10.2) differs little from that of fixed boundary value problems with one nonlinear boundary condition. We again make use of the affine transformation

$$u(t) = U(t)x(t) + v(t)$$

where U and v satisfy the initial value problems (2.6.2) and (2.6.3) and may, at least in principle, be considered known for all $t \geq 0$. The first of the boundary conditions of (2.10.2) requires that $x(t)$ be chosen such that

$$g(U(t)x + v(t), x) = 0 \qquad (2.10.3)$$

We shall assume that this equation can be solved for x as a function of t. Given the solution $x(t)$ of (2.10.3), it remains to find a T such that

$$h(U(t)x(t) + v(t), x(t)) = 0 \qquad (2.10.4)$$

for $t = T$. Note that (2.10.4) is a scalar equation in t and, at least from a numerical view, is readily solvable. Once T is known, the boundary values $(U(T)x(T) + v(T), x(T))$ are consistent with the given boundary values of (2.10.1), and (2.10.1) is reduced to a standard initial value problem.

It is possible to modify the existence theorems of Section 2.3 in order to obtain existence and uniqueness theorems for the free boundary value problem (2.10.1) and (2.10.2). Two sets of hypotheses are required. The first set will guarantee that the initial value problem (2.6.2) for the Riccati equation has a solution over an interval $[0, \hat{T})$. Theorem 1.1.3, for example, will apply in this case. The second set will assure that the boundary conditions (2.10.2) can be satisfied for some $T \in [0, \hat{T})$. To give general existence criteria will be rather cumbersome, and it may be more instructive to consider some examples.

Example 2.10.1. A Minimum Fuel Problem

An unconstrained minimum fuel problem for a controlled linear dynamical system may be expressed as

$$J = S(T, x(T)) + \int_0^T [\langle Q(t)x(t), x(t)\rangle + \langle R(t)u(t), u(t)\rangle]\, dt$$

$$x' = A(t)x + B(t)u, \qquad x(0) = x_0, \qquad x(T) = x_1 \qquad (2.10.5)$$

The object is to find a differentiable trajectory $x(t)$ and control $u(t)$, and a final time T, which yield a minimum for the cost functional J.

We shall formally present an algorithm for the solution of (2.10.5) which attacks the free boundary value problem derived from (2.10.5) through the use of Lagrange multipliers. It is shown in detail in Sage (1968, p. 61) that the optimal solution of (2.10.5) necessarily has to satisfy the equations

$$
\begin{aligned}
x' &= A(t)x - B(t)R^{-1}(t)B^{\mathrm{T}}(t)\lambda, & x(0) &= x_0 \\
\lambda' &= -Q(t)x - A^{\mathrm{T}}(t)\lambda, & x(T) &= x_1
\end{aligned}
\tag{2.10.6}
$$

$$
\frac{\partial S}{\partial T}(T, x_1) + \langle Q(T)x_1, x_1 \rangle + \langle A(T)x_1 - B(T)R^{-1}(T)B(T)\lambda(T), \lambda(T) \rangle = 0
\tag{2.10.7}
$$

with the optimal control given by $u(t) = -R^{-1}(t)B^{\mathrm{T}}(t)\lambda(t)$. Equation (2.10.7) is the scalar equation used to determine the free boundary T. For its solution we need a representation for $\lambda(T)$. To this end, invariant imbedding is applied. We know from the discussion of linear boundary value problems that

$$
x(t) = U(t)\lambda(t) + v(t)
$$

where U and v are the solutions of (2.6.2) and (2.6.3), in this case

$$
\begin{aligned}
U' &= -B(t)R^{-1}(t)B^{\mathrm{T}}(t) + A(t)U + UA^{\mathrm{T}}(t) + UQ(t)U, & U(0) &= 0 \\
v' &= [U(t)Q(t) + A(t)]v, & v(0) &= x_0
\end{aligned}
$$

To satisfy the boundary condition $x(t) = x_1$ at an arbitrary point t we see that $\lambda(t)$ must be chosen as

$$
\lambda(t) = U^{-1}(t)[x_1 - v(t)]
$$

Substitution into (2.10.7) shows that T must be a root of the equation

$$
\begin{aligned}
F(t) &\equiv (\partial S/\partial t)(t, x_1) + \langle Q(t)x_1, x_1 \rangle + \langle A(t)x_1 - B(t)R^{-1}(t)B^{\mathrm{T}}(t)U^{-1}(t) \\
&\quad \times [x_1 - v(t)], \; U^{-1}(t)[x_1 - v(t)] \rangle = 0
\end{aligned}
\tag{2.10.8}
$$

In principle U and v are known, hence F is known and the problem reduces to finding the zero of the scalar function (2.10.8). This problem is generally easy to handle numerically with, for example, a search method.

We note that if R and Q are bounded and positive definite on $(0, \infty)$, then $U(t)$ exists and is positive definite according to Corollary 1.3.1. Hence $\lambda(t)$ is defined (see N.10.2). Conditions which assure the uniqueness of a root for (2.10.8) then become sufficient for the uniqueness of the solution of (2.10.6) and (2.10.7). Whether this solution actually yields the minimum of J must be determined independently because the Lagrange multiplier formulation only yields a necessary condition for optimality.

As an illustration, consider the scalar problem presented in Sage (1968, p. 63).

$$x' = u, \qquad x(0) = 1, \qquad x(T) = 0$$

$$J = T + \tfrac{1}{2}\beta \int_0^T u^2 \, dt$$

We see that in this case $A = Q = 0$, $B = 1$, $R = \tfrac{1}{2}\beta$, $S = T$. The Riccati equation assumes the form

$$U' = -2/\beta, \quad U(0) = 0 \quad \text{thus} \quad U(t) = -2t/\beta$$

The linear equation for v has the constant solution $v(t) = 1$. The scalar equation (2.10.8) is given by

$$F(t) \equiv 1 - \beta/2t^2 = 0, \qquad \text{hence} \quad T = (\beta/2)^{1/2}$$

The function $\lambda(t)$ is given by

$$\lambda' = -[Q(t)U(t) + A^T(t)]\lambda - Q(t)v(t), \quad \lambda(T) = U^{-1}(T)[x_1 - v(T)]$$

(see Eq. (2.6.5)), which in this case has the solution

$$\lambda(t) = \beta/2T = (\beta/2)^{1/2}$$

Hence, the control is given as $u(t) = -(\tfrac{1}{2}\beta)^{-1}\lambda(t) = -(2/\beta)^{1/2}$, while the trajectory is

$$x(t) = 1 - (2/\beta)^{1/2}t \quad \blacksquare$$

Example 2.10.2. The One-Phase Stefan Problem

As a second example, let us consider a free boundary value problem for the one-dimensional diffusion equation, a so-called Stefan problem. The melting of ice, fluidized-bed coating, drying of a vertical porous column, the filtration of liquids, and the process of chemical precipitation can often be modeled by the following one-dimensional, one-phase Stefan problem

$$\partial^2 u/\partial x^2 + B(t, x)\,(\partial u/\partial x) - C(t, x)u - D(t, x)\,(\partial u/\partial t) = F(t, x)$$

$$u(0, x) = u_0(x), \qquad x \in [0, s_0], \quad s_0 \geq 0$$

(2.10.9)

subject to the fixed boundary condition

$$a(t)u(t, 0) = b(t) \frac{\partial u}{\partial x} (t, 0) + c(t) \tag{2.10.10}$$

and the free boundary conditions at $x = s(t)$

$$\begin{pmatrix} d(t, s(t), s'(t)) & e(t, s(t), s'(t)) \\ f(t, s(t), s'(t)) & g(t, s(t), s'(t)) \end{pmatrix} \begin{pmatrix} u(t, s(t)) \\ \frac{\partial u}{\partial x} (t, s(t)) \end{pmatrix} = \begin{pmatrix} h(t, s(t), s'(t)) \\ k(t, s(t), s'(t)) \end{pmatrix}$$

$$\tag{2.10.11}$$

A solution of (2.10.9)–(2.10.11) is a pair of functions $\{u(t, x), s(t)\}$ such that u is a solution of (2.10.9) and (2.10.10) on $0 \le x \le s(t)$, $t > 0$ and such that $u(t, s(t))$ and $s(t)$ satisfy Eqs. (2.10.11) (see N.10.3).

This formulation is quite general and for many problems reduces to a much simpler expression. For instance, the melting of a slab of ice subject to a variable temperature input on the fixed boundary allows us to set $B \equiv C \equiv F \equiv b \equiv e \equiv f \equiv h \equiv 0$, $D(t, x) \equiv D$, $a \equiv 1$, $d \equiv 1$, $g \equiv 1$, $k(t, s(t), s'(t)) = ks'(t)$, so that the complete formulation is

$$\partial^2 u / \partial x^2 - \partial u / \partial t = 0$$

$$u(0, x) = u_0, \qquad u(t, 0) = c(t), \qquad u(t, s(t)) = 0, \qquad \frac{\partial u}{\partial x} (t, s(t)) = ks'(t)$$

The simplest way of converting the Stefan problem into a free boundary value problem for ordinary differential equations is to replace the time derivatives by their backward differences. Choosing a constant time step Δt and writing $B_n(x) \equiv B(n \Delta t, x)$, $n = 1, 2, \ldots$, etc., we can approximate (2.10.9)–(2.10.11) by the first-order system

$$u_n' = y_n$$
$$y_n' = [C_n(x) + D_n(x)/\Delta t]u_n - B_n(x)y_n + F_n(x) - (D_n(x)/\Delta t)u_{n-1}(x)$$
$$a_n u_n(0) = b_n y_n(0) + c_n$$

$$\tag{2.10.12}$$

$$\begin{pmatrix} d_n\left(s, \dfrac{s - s_{n-1}}{\Delta t}\right) & e_n\left(s, \dfrac{s - s_{n-1}}{\Delta t}\right) \\ f_n\left(s, \dfrac{s - s_{n-1}}{\Delta t}\right) & g_n\left(s, \dfrac{s - s_{n-1}}{\Delta t}\right) \end{pmatrix} \begin{pmatrix} u_n(s) \\ y_n(s) \end{pmatrix} = \begin{pmatrix} h_n\left(s, \dfrac{s - s_{n-1}}{\Delta t}\right) \\ k_n\left(s, \dfrac{s - s_{n-1}}{\Delta t}\right) \end{pmatrix}$$

If the temperature (concentration, etc.) u is given at $x = 0$, then $a = 1$ and we shall consider y as the base characteristic; if flux is given, then $b = 1$

and u is the base characteristic when applying invariant imbedding to (2.10.12). For definiteness suppose that $a = 1$. Then the general theory states that at the unknown boundary $x = s$

$$u_n(s) = U_n(s)y_n(s) + v_n(s) \qquad (2.10.13)$$

where U and v satisfy the initial value problems (2.6.2) and (2.6.3), respectively, which in this case assume the form

$$U_n' = 1 + B_n(x)U - [C_n(x) + D_n(x)/\Delta t]U_n^2, \qquad U_n(0) = b_n$$
$$v_n' = -U_n(x)[C_n(x) + D_n(x)/\Delta t]v_n - U_n(x) \qquad (2.10.14)$$
$$\times [F_n(x) - (D_n(x)/\Delta t)u_{n-1}(x)], \qquad v_n(0) = c_n$$

Substitution of (2.10.13) into the free boundary condition yields two linear equations in $y_n(s)$ which allows us to eliminate $y_n(s)$. Remaining is a single equation involving only the coefficients, U_n and v_n, and the unknown s. In fact, it is straightforward to verify that this equation is given by

$$F_n(s) \equiv (h_n - d_n v_n)(f_n U_n + g_n) - (k_n - f_n v_n)(d_n U_n + e_n) = 0 \quad (2.10.15)$$

where the arguments of h_n, etc., have been suppressed for ease of notation. As the equations for U and v are integrated, we can evaluate F_n; if $F_n(s) = 0$ for some $s = s_n$, then s_n is an admissible free boundary at the time level $t = n\,\Delta t$ and the complete solution can be found by integrating over $[0, s_n]$ either (2.10.12) subject to

$$u_n(s_n) = U_n(s_n)y_n(s_n) + v_n(s_n)$$
$$y_n(s_n) = (h_n - d_n v_n)/(d_n U_n + e_n) \qquad (2.10.16)$$

or $(y_n(s_n) = (k_n - f_n v_n)/(f_n U_n + g_n))$, or by solving the backward sweep equation (2.2.5), which in this case is

$$y_n' = [C_n(x) + D_n(x)/\Delta t][U_n(x)y_n + v_n]$$
$$- B_n(x)y_n + F_n(x) - (D_n(x)/\Delta t)u_{n-1}(x) \qquad (2.10.17)$$

with $y_n(s)$ given as above, and by setting

$$u_n(x) = U_n(x)y_n(x) + v_n(x)$$

In connection with classical Stefan problems and monotone boundaries this approach has been studied by Meyer (1970c). It is shown that the Riccati equation has a unique bounded solution on $[0, \infty)$, that $F(s) = 0$

has a unique solution, and that u_n converges to the unique solution $u(n \, \Delta t, x)$ of the Stefan problem as $\Delta t \to 0$. In addition several numerical examples were presented which are relevant to our discussion.

The first of these was the Stefan problem for the heat equation which describes the drying of a vertical porous column. The equations were

$$\partial^2 u/\partial x^2 - D \, \partial u/\partial t = 0$$

$$u(t, 0) = c, \qquad u(t, s(t)) = 0, \qquad \frac{ds}{dt} = -\gamma \, \frac{\partial u}{\partial x} \, (t, s(t)), \qquad s(0) = 0$$

where $D = 1.225$, $c = 0.175$, and $\gamma = 0.000643$. This Stefan problem was solved by integrating Eqs. (2.10.14) (note that U is independent of time and has to be found only once), finding the root of (2.10.15), and then integrating backward the initial value problem (2.10.12) subject to the initial values (2.10.16).

The integrations were carried out with a fourth-order Runge–Kutta method using time and space steps of $\Delta t = 100$ and $\Delta x = 10^{-2}$. Figure 2.10.1 shows a plot of s^2 as a function of time. The method performed well and was not affected by discontinuity of the data at $t = 0$.

The same approach was taken to solve the Stefan problem

$$\partial^2 u/\partial x^2 - \partial u/\partial t = F(t, x) \tag{2.10.18}$$

$$u(t, 0) = c(t), \qquad u(t, s(t)) = 0, \qquad \frac{ds}{dt} + \frac{\partial u}{\partial x} \, (t, s(t)) = \mu(t, x)$$

Fig. 2.10.1. Plot of time versus s^2 for a drying porous column [from Meyer (1970c)].

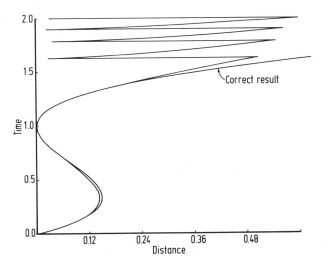

Fig. 2.10.2. Unstable computation using back integration of the characteristic equations [from Meyer (1970c)].

where the data functions were computed such that this problem had the solution $\{t(1 - t)^2 - x, t(1 - t)^2\}$. With $\Delta t = \Delta x = 0.02$, no problem was encountered initially in finding $s(t)$ and $u(t, x)$. However, the method became unstable. Figure 2.10.2 shows a typical computer plot of the free boundary. Changing space and time steps did not help. This instability

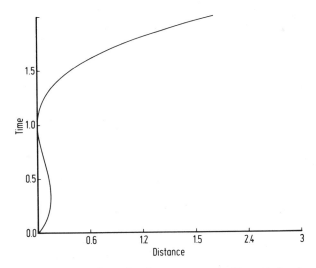

Fig. 2.10.3. Stable computation using the reverse sweep. Computed and exact solution coincide [from Meyer (1970c)].

is due to the exponential growth of errors introduced into the calculation. On the other hand, when u_n was determined from (2.10.17) and (2.10.13), Fig. 2.10.3 was obtained. This example points out the usefulness of invariant imbedding for problems where the shooting method is likely to fail. ∎

The application of invariant imbedding to problems with a free boundary is attractive for linear characteristic differential equations and a linear initial manifold because the surface characteristic $u(t)$ and base characteristic $y(t)$ are related through the affine transformation

$$u(t) = U(t)y(t) + v(t)$$

where, at least in principle, U and v are known matrices and functions. If nonlinear problems are to be solved, this simple structure is lost. However, it remains useful to consider the surface $u(t, x)$ as being generated with the shooting method applied to the characteristic equations and to search for those points on u which satisfy the boundary conditions. Once those points are found, the corresponding characteristic is the solution of the free boundary value problem. Specifically, consider the system

$$y'(t) = F(t, u, x), \qquad\qquad u(0) = f(x(0))$$
$$u'(t) = G(t, u, x), \qquad g(u(T), x(T)) = 0$$

(2.10.19)

where T has to be determined such that the scalar equation

$$h(u(T), x(T)) = 0$$

holds. It follows from Section 2.2 that the solution of (2.10.19) must pass through the locus

$$C = \{(t, x): g(u(t, x), x) = 0\}$$

where u is the integral surface of (2.2.6) because a characteristic through C necessarily satisfies the given boundary conditions. If this locus can be found explicitly as $x = x(t)$, then the free boundary $t = T$ is the root of the equation

$$F(t) \equiv h(g(u(t, x(t)), x(t)) = 0$$

Let us illustrate this general algorithm with the following hypothetical example.

Example 2.10.3. A Nonlinear Dynamical System

Suppose a particle is propelled from the origin $u = 0$ to a final position $u = 1$ against a force given as $-1 - u - u'^2$. It is desired to determine the initial velocity $u'(0)$ which assures that the particle is momentarily at rest at $u = 1$, or equivalently, the duration of motion T.

The formulation of this problem is

$$u'' = -1 - u - u'^2, \quad u(0) = 0, \quad u(T) = 1, \quad u'(T) = 0$$

or in first-order form

$$u' = x, \qquad\qquad u(0) = 0$$
$$x' = -1 - u - x^2, \quad u(T) = 1 \qquad (2.10.20)$$
$$x(T) = 0$$

where the last equation is the scalar equation necessary to determine the free boundary T. The integral surface $u(t, x)$ corresponding to (2.10.20) was found numerically over a triangle with corners at $(t, x) = (0, -1.5)$,

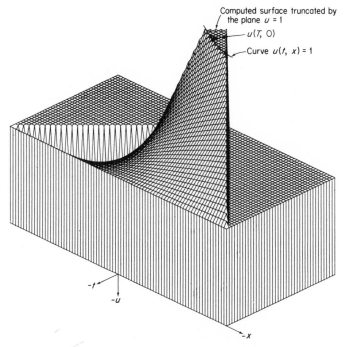

Computed surface truncated by the plane $u = 1$

$u(T, 0)$

Curve $u(t, x) = 1$

$-t$

$-u$

$-x$

Fig. 2.10.4. Geometry of the free interface locus for the dynamical problem.

(0, 1.5), (1.5, 0) with the conservation method of Lax (see Section 2.4) and is shown in Fig. 2.10.4. The intersection with the surface $u = 1$ is also indicated. This is a simple curve, and where it crosses the axis $x = 0$ the desired boundary is located. The characteristic through this point is the solution of (2.10.20). As a check, the computed value of $T = 1.2651$ was used as the final time for a backward integration of (2.10.20) with a fourth-order Runge–Kutta method. The resulting solution $u(t)$ is graphed in Fig. 2.10.5. A value of $u(0) = 0.0163$ (instead of 0) was found when a mesh size of $\Delta t = \Delta x = 3/80$ was used. This result was considered acceptable and no further refinement of the mesh was attempted. Incidentally, the computation was carried through with this reasonably course mesh to allow easy plotting of the surface. ∎

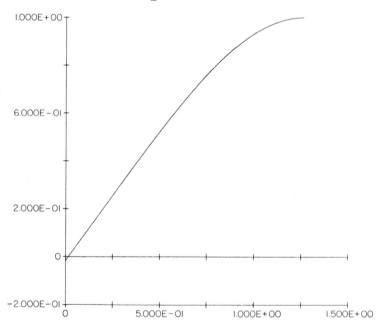

Fig. 2.10.5. Back integration of the equations of motion from the computed free boundary to the origin.

The technical difficulties of solving free boundary value problems increase rapidly with the dimension of the base characteristic. For many problems trial and error or iteration methods may well be preferable. Possibly, a crude integration of the invariant imbedding equation may furnish a good initial guess for iterative methods. Finally, it appears possible to extend this analysis to problems where both endpoints are to be determined as part of the solution. The major difficulty now is due to the observation that even for linear characteristic equations the free boundary value problem is inherently nonlinear. For linear equations with one nonlinear boundary condition or one free boundary condition, the treatment of the nonlinearity could be postponed until after the integral surface was found. Nonlinear boundary values or two free boundaries destroy the connection between surface and base characteristic as defined by the affine transformation $u(t) = U(t)y(t) + v(t)$.

2.11. DISCRETE TWO-POINT BOUNDARY VALUE PROBLEMS

Analogously to the continuous case, the discrete characteristic theory of Section 1.5 lends itself to the conversion of boundary value problems for difference equations into a sequence of initial value problems. Discrete problems may arise through the discretization of continuous boundary value problems for ordinary and partial differential equations or they may describe inherently discrete processes such as time discrete control systems. The subject of discrete systems in the theory of optimal control is far ranging and well beyond the scope of this book. A brief introductory discussion and additional references on this topic as well as an application of boundary perturbation techniques to discrete boundary value problems may be found in the monograph of Sage and Melsa (1971) on systems identification. In our volume the emphasis is on differential equations and we shall only indicate how characteristic theory may be applied to difference equations.

Without regard to the origin of the equations, we shall assume that we are faced with a discrete two point boundary value problem of the form

$$u(k + 1) = F(k, u(k), x(k)), \qquad u(0) = f(x(0))$$
$$x(k + 1) = G(k, u(k), x(k)), \qquad g(u(N), x(N)) = 0 \qquad (2.11.1)$$

for $k = 0, \ldots, N - 1$. Here we shall assume that u and x are m- and n-dimensional vectors, respectively. Note that in this problem the parameter t is discretized. The vector $x(k)$ can assume all values in E^n. Our aim is now to derive a sequence of initial value problems to determine the solution $\{u(k), x(k)\}$, $k = 0, \ldots, N$.

As in the continuous case the boundary value problem is imbedded into the family of initial value problems,

$$u(k + 1) = F(k, u(k), x(k)), \qquad u(0) = f(s)$$
$$x(k + 1) = G(k, u(k), x(k)), \qquad x(0) = s \qquad (2.11.2)$$

where s is an arbitrary n-dimensional vector. The comments following Theorem 1.5.1 indicate that as s ranges (continuously) over E^n the corresponding solutions $\{u(k, s), x(k, s)\}$ generate the time discrete "integral surface" $u(k, x(k))$ of the corresponding discrete invariant imbedding equation

$$R(k + 1, G(k, R(k, x(k)), x(k))) = F(k, R(k, x(k)), x(k))$$

$$R(0, x(0)) = f(x(0))$$

This equation can be solved recursively for $k = 1, \ldots, N$. Suppose now that $R(N, x(N))$ has been found for all $x(N)$ in a sufficiently large domain, and suppose that there exists a vector $\hat{x}$ such that $g(R(N, \hat{x}), \hat{x}) = 0$. Then it follows from Theorem 1.5.1 that the discrete characteristic $\{u(k), x(k)\}$ through the point $(R(N, \hat{x}), \hat{x})$ at $k = N$ is indeed the solution of the boundary value problem (2.11.1).

There are some notational and conceptual difficulties with this formulation which will bear further comment. First of all, $x(k)$ is the independent variable at the kth step and consists of an n-dimensional vector ranging over E^n. However, $R(k, x(k))$ may not be defined for all $x(k)$; indeed, $R(1, x(1))$ is defined only for those $x(1)$ in the range R_1 of $G(0, f(x(0)), x(0))$ as $x(0)$ ranges over $R_0 = E^n$. At Step 2, $R(2, x(2))$ is defined only for all $x(2)$ belonging to the range R_2 of $G(1, R(1, x(1))x(1))$ as $x(1)$ ranges over R_1. The ranges R_n, $n = 1, \ldots, N$ may shrink or even become empty, so that a simple boundary condition like $x(N) = b$ may be impossible to satisfy. This brief discussion will indicate that the analytical difficulties inherent in the discrete formulation cannot be neglected.

From a numerical point of view it would appear that a practical algorithm for the solution of (2.11.2) actually corresponds to the shooting method. We start with K initial values $\{x_l(0)\}_{l=1}^{l=K}$, determine the corresponding $\{x_l(1)\}_{l=1}^{K}$ and $\{R(1, x_l(1))\}_{l=1}^{K}$ and proceed to the next step. A solution $R(n, x(n))$ over a constant domain of x for all n would likely require considerable inter- and extrapolation and thus not be feasible. However, no representative examples are offered in this regard.

The difficulties largely vanish when the problem is linear. Specifically, consider the difference equations

$$u(k + 1) = A(k)u(k) + B(k)x(k) + F(k), \qquad u(0) = fx(0) + a$$
$$x(k + 1) = C(k)u(k) + D(k)x(k) + G(k), \qquad g(u(N), x(N)) = 0 \tag{2.11.3}$$

It was established in Section 1.5 that $u(k)$ and $x(k)$ are related through the affine transformation

$$u(k) = U(k)x(k) + w(k)$$

where U and w are computed from the forward sweep

$$U(k + 1)(C(k)U(k) + D(k)) - A(k)U(k) - B(k) = 0, \qquad U(0) = f$$

and

$$w(k + 1) + U(k + 1)(C(k)w(k) + G(k)) - A(k)w(k) - F(k) = 0,$$
$$w(0) = 0$$

At $k = N$ we now have to find a root $\hat{x}$ of the equation

$$g(U(N)x + w(N), x) = 0$$

If such $\hat{x}$ exists then the complete solution $\{u(k), x(k)\}$ can be computed from the backward sweep for $x(k)$, namely

$$(C(k)U(k) + D(k))x(k) + C(k)w(k) + G(k) - x(k + 1) = 0, \quad x(N) = \hat{x}$$

and by setting

$$u(k) = U(k)x(k) + w(k)$$

Alternatively, we can solve the difference equations (2.11.1) backward from $k = N$ to $k = 0$ subject to the initial values

$$u(N) = U(N)\hat{x} + w(N), \qquad x(N) = \hat{x}$$

These forward and backward sweeps are, of course, analogous to the sweeps for continuous systems, and as in the continuous case it is possible to verify the validity of the solutions $u(k) = U(k)x(k) + w(k)$, $x(k)$ through mechanical manipulations, in this case direct substitution rather than differentiation and substitution. Indeed, from the affine transformation and the defining equations for U and w we obtain

$$
\begin{aligned}
u(k + 1) &= U(k + 1)x(k + 1) + w(k + 1) \\
&= U(k + 1)(C(k)U(k) + D(k))x(k) - U(k + 1)(C(k)w(k) \\
&\quad + G(k)) + F(k) + A(k)w(k) - U(k + 1)(C(k)w(k) + G(k)) \\
&= A(k)U(k)x(k) + B(k)x(k) + A(k)w(k) + F(k) \\
&= A(k)u(k) + B(k)x(k) + F(k)
\end{aligned}
$$

The boundary values are satisfied by construction so that the $\{u(k), x(k)\}$ constructed with the sweeps is a solution of (2.11.1).

For the numerical advantage of solving difference equations obtained from continuous boundary value problems for elliptic equations with the sweep method we refer to Berezin and Zhidkov (1965) (N.11.1).

NOTES

N.1.1. Some partial results are available in this regard. Roberts and Shipman (1970) have carried through a comparison between the method of adjoints and complementary functions. They show that both methods lead to the same algebraic system for the determination of the missing boundary values although the methods of computation are dissimilar. Stability considerations determine which method is preferable for a given problem. The relation between the methods of particular integrals and complementary functions is treated in Miele (1970). The equivalence between the sweep method (see Gelfand and Fomin (1963)) and the method of invariant imbedding for linear problems is established in Meyer (1968).

N.1.2. An early but very complete list of references on invariant imbedding, particularly with regard to radiative transfer, may be found in the monograph of Wing (1962) on transport theory. A computer based updated bibliography on invariant imbedding is being maintained by M. R. Scott (1972).

N.1.3. It is recognized that considerable generality is lost by reformulating (2.2.11) as a problem with separated boundary conditions. A more general imbedding involving some characteristic parameters such as the length of the interval of integration may well be preferable for a given problem. This approach is explored by Keller (1972). However, the geometric interpretation of the resulting equations via characteristic theory will usually be lost.

N.4.1. For ease of notation we shall speak of the independent variable t as time. In a concrete application involving time but not as independent variable we shall generally use ξ as independent variable.

N.4.2. A recent study of high order conservation-type approximations for related equations has been presented by Zwas and Abarbanel (1971). However, they generally require that $F = 0$, so that their results appear to be of limited value for our problems.

N.5.1. The author is indebted to Y. Ozawa of the Mobil Research and Development Corporation for pointing out this example.

N.5.2. Invariant imbedding has found frequent application to chemical engineering problems. The book of Lee (1968) contains several examples involving chemical reactors, as well as additional references.

N.5.3. The Thomas-Fermi equation has been studied extensively. We refer to the physics paper of March (1952) in which the Thomas–Fermi equation is solved numerically with the shooting method, and to the mathematics paper of Mason (1966) where rational approximations are used. A complete discussion of the analytical aspects of this equation may be found in the book of Bellman (1953).

Added in proof: A moment's reflection should have made it apparent to us that integrating (2.5.6), subject to $u(T, x) = Tx$, backward over $[0, T]$ would have been more reasonable because the characteristic through $x(T) = u(T) = 0$ coincides with the ξ-axis. The method of Shampine and Thompson can now be used.

N.7.1. The application of invariant imbedding to a hybrid integration of the one-dimensional diffusion equation has been discussed by Nelson and Altom (1971). It is well known that this problem requires the forward–backward sweep approach of invariant imbedding (or equivalent methods) to overcome the instability inherent in the shooting approach to solving the diffusion equation.

N.8.1. Gourlay (1970) has pointed out that for rapidly decaying functions the trapezoidal rule may cease to be locally stable above a certain mesh size. The problem is eliminated if $u' = F(t, u)$ is integrated with the modified trapezoidal rule

$$u_{n+1} - u_n = hF(t_{n+1/2}, (u_{n+1} + u_n)/2)$$

We were not aware in time of Gourlay's work and have used only the standard trapezoidal rule

$$u_{n+1} - u_n = h(F(t_{n+1}, u_{n+1}) + F(t_n, u_n))/2$$

However, no difficulties were observed.

N.10.1. This representation was chosen for ease of comparison with published results for the linear regulator. For numerical work the inverse representation $\lambda(t) = W(t)x(t) + r(t)$ would appear to be more convenient.

N.10.2. Not only are the invariant imbedding equations well defined in this case, but invariant imbedding likely is superior to other commonly used solution methods. Spingarn (1971) has compared the performance of invariant imbedding for linear regulator problems (with fixed final time t_f) with the matrix Riccati method (which is identical with the method of invariant imbedding when Step 4′ is used) and the method of complementary integrals. Because of superior stability invariant imbedding gave more accurate results.

N.10.3. Stefan problems have received considerable attention in the literature. For a recent reference on the analytical aspects the reader is referred to Sherman (1971). Numerical aspects and further references on the one phase case are found in Meyer (1970c).

N.11.1. It would be interesting to examine whether the unconventional finite difference schemes of Bellman and Cooke (1965) and Lee (1968) for the continuous invariant imbedding equation do indeed correspond to a discretization of the underlying boundary value problem for ordinary differential equations. However, such an analysis has not yet been carried through.

Chapter

3

Interface Problems

3.1. THE INVARIANT IMBEDDING EQUATIONS FOR INTERFACE PROBLEMS

In many physical systems modeled by differential equations, the parameters of the system may not be continuous functions of time and space. Often these discontinuities are reflected in interface conditions for the differential equations. For example, heat conduction in a composite slab extending from $\xi = 0$ to $\xi = X$ with an interface at $\xi = L$ can be described by the heat equation

$$\partial/\partial\xi(k_1\partial u/\partial\xi) - \varrho_1 c_1(\partial u/\partial t) = 0 \quad \text{on} \quad 0 < \xi < L$$

$$\partial/\partial\xi(k_2\partial u/\partial\xi) - \varrho_2 c_2(\partial u/\partial t) = 0 \quad \text{on} \quad L < \xi < X$$

subject to temperature or flux conditions at $\xi = 0$ and $\xi = X$, while at the interface $\xi = L$ one requires continuity of temperature and heat flux, which yield the interface conditions

$$u(t, L - \varepsilon) = u(t, L + \varepsilon)$$

$$k_1 \frac{\partial u}{\partial x}(t, L - \varepsilon) = k_2 \frac{\partial u}{\partial x}(t, L + \varepsilon)$$

as $\varepsilon \to 0$. The material properties (k, ϱ, c) generally will be different for each slab. In addition, they may be temperature dependent so that the heat equation and the interface conditions actually are nonlinear. Several examples of fixed and free interface problems will be solved in subsequent sections.

Numerical techniques for the solution of two-point boundary value problems for ordinary differential equations are generally applicable to fixed interface problems as well. For instance, in the shooting method one-sided limits at the interface are known so that the solution can be continued past the interface by computing new initial values from the interface condition. In finite difference techniques the algebraic equation at the interface mesh point is derived from the interface condition rather than the differential equations while in Galerkin techniques, when applicable, the interface condition is taken into account through the choice of the base functions (see N.1.1). Equally applicable is invariant imbedding; however, the modifications required for the invariant imbedding equations are not as evident and will therefore be derived in detail. Again, the characteristic theory approach will prove fruitful.

Consider the general nonlinear two-point boundary value problem

$$u_i{}' = F_i(t, u_i, x_i),$$
$$x_i{}' = G_i(t, u_i, x_i), \qquad i = 1, 2 \qquad (3.1.1)$$

where u_1 and x_1 are m- and n-dimensional vectors, respectively, defined on $[0, L]$, and where u_2, x_2, etc., are defined on $[L, T]$. We shall assume that Eqs. (3.1.1) are subject to the boundary conditions

$$u_1(0) = f(x_1(0))$$
$$g(u_2(T), x_2(T)) = 0 \qquad (3.1.2)$$

where $f = (f_1, \ldots, f_m)$ and $g = (g_1, \ldots, g_n)$ are given functions, while at the interface $t = L$ we impose the interface condition

$$H(u_1(L), u_2(L), x_1(L), x_2(L)) = 0 \qquad (3.1.3)$$

Note that H must be an $(m + n)$-dimensional vector valued function if all components of (3.1.1) are to be accounted for at the interface. For definiteness we shall assume that all given functions are continuously differentiable.

We proceed as in Section 2.2 and imbed the given boundary value problem into the family of initial value problems

$$u_1{}' = F_1(t, u_1, x_1), \qquad u_1(0) = f(s)$$
$$x_1{}' = G_1(t, u_1, x_1), \qquad x_1(0) = s$$

where $s = (s_1, \ldots, s_n)^T$ is a free parameter ranging over E^n. If the solution $\{u_1(t, s), x_1(t, s)\}$ is considered a characteristic, and if the inverse function $s = s(t, x)$ exists, then it follows from Section 1.2 that $u_1(t, s(t, x))$ satisfies the first-order partial differential equation

$$u_{1t} + u_{1x} G_1(t, u_1, x_1) = F_1(t, u_1, x_1)$$
$$u_1(0, x_1) = f(x_1) \tag{3.1.4}$$

Consequently, the solution of (3.1.1)–(3.1.3) is also imbedded into the interface condition

$$H(u_1(L, x_1), x_1, u_2, x_2) = 0 \tag{3.1.5}$$

Let us now suppose that the $m + n$ equations (3.1.5) admit a solution of the form

$$u_2 = h(u_1(L, x_1), x_2)$$
$$x_1(L) = k(x_2) \tag{3.1.6}$$

where k is an invertible mapping (homeomorphism) over E^n. As x_2 varies over E^n, (3.1.6) describes an initial manifold

$$u_2(L, x_2) = h(u_1(L, k(x_2), x_2) \tag{3.1.7}$$

and the characteristics through this manifold for $t > L$ form the integral surface u_2 of

$$u_{2t} + u_{2x} G_2(t, u_2, x_2) = F_2(t, u_2, x_2) \tag{3.1.8}$$

Finally, if the equation

$$g(u_2(T, x), x) = 0 \tag{3.1.9}$$

has a root $\hat{x}_2$, then the initial value $(u_2(T, \hat{x}_2), \hat{x}_2)$ is consistent with the boundary and interface conditions (3.1.2) and (3.1.3). In summary, we can state the following theorem which is our invariant imbedding recipe for interface problems.

THEOREM 3.1.1. Assume that

(i)   the invariant imbedding equation

$$u_{1t} + u_{1x} G_1(t, u_1, x_1) = F_1(t, u_1, x_1)$$
$$u_1(0, x_1) = f(x_1) \tag{3.1.10}$$

has a solution $u_1(t, x)$ over $[0, L] \times E^n$;

(ii)   the equation $H(u_1(L, x_1), u_2, x_1, x_2) = 0$ has a solution

$$u_2 = h(x_2), \qquad x_1 = k(x_2) \tag{3.1.11}$$

for all $x_2 \in E^n$;

(iii)   the invariant imbedding equation

$$u_{2t} + u_{2x} G_2(t, u_2, x_2) = F_2(t, u_2, x_2)$$
$$u_2(L, x_2) = h(x_2) \tag{3.1.12}$$

has a solution $u_2(t, x_2)$ over $[L, T] \times E^n$;

(iv)   the equation

$$g(u_2(T, x), x) = 0 \tag{3.1.13}$$

has a root $\hat{x}_2$. Then the characteristic $\{u_2(t), x_2(t)\}$ through $(u_2(T, \hat{x}_2), \hat{x}_2)$ is a solution of (3.1.1)–(3.1.3) over $[L, T]$, while the characteristic $\{u_1(t), x_1(t)\}$ through $(u_1(L, k(x_2(L)), x_2(L))$ is the solution over $[0, L]$.

Proof. By hypothesis $\{u_2, x_2\}$ satisfy the boundary condition at $t = T$. Since the characteristic remains imbedded in the integral surface, it passes through $u_2(L, x_2) = h(x_2)$. By hypothesis the point $(u_1(L), x_1(L))$ defined by $x_2(L)$ satisfies the interface condition, while the corresponding characteristic through $(u_1(L), x_1(L))$ remains imbedded in $u_1(t, x_1)$ and thus fulfills the boundary condition at $t = 0$. ∎

It is apparent that this theory can be extended to multiple interfaces. In fact, given a sequence of interfaces located at $0 < L_1 < \cdots < L_N < T$ and interface conditions $\{H_i\}_{i=1}^N$, we start as above, except that at $t = L_i$ we go through steps (ii) and (iii) in Theorem 3.1.1 repeatedly until we reach the fixed boundary at $t = T$. If, however, there is only one interface condition to be satisfied and if the boundary condition at $t = T$ can be solved for $x_2(T) = g(u_2(T))$, then it may be advantageous to integrate (3.1.4) over $[0, L]$ and

$$x_{2t} + x_{2u_2}F(t, u_2, x_2) = G(t, u_2, x_2)$$
$$x_2(T, u_2) = g(u_2)$$

backward over $[L, T]$. The solutions $u_1(L, x_1)$ and $x_2(L, u_2)$ are then substituted into the interface condition (3.1.3). We now have $m + n$ equations in the $m + n$ unknowns x_1, u_2. If (3.1.3) can be solved, the characteristics through $(u_1(L, x_1), x_1)$ and $(u_2, x_2(L, u_2))$ are the solution of (3.1.1)–(3.1.3).

The problem of actually determining the characteristic curves once the integral surfaces are known can be attacked either by integrating the differential equations (3.1.1) subject to the initial value found with the invariant imbedding technique, or the reverse sweep based on equation (2.2.8) may be used. The reverse sweep should be attempted when the Cauchy problem for (3.1.1) is unstable.

It is possible to derive existence theorems for interface problems with the invariant imbedding approach. In fact, all that is needed is an estimate for $\| u_{2x_2} \|$ along the interface $t = L$ in order to continue the surface past L. This problem is little different from solving the boundary equation $g(u(T, x), x) = 0$ in Section 2.3, and it is not difficult to prove results analogous to Theorems 2.3.1–2.3.4 for the above interface problem. The conclusion generally is that for sufficiently simple interface and boundary conditions the problem has a unique solution provided the intervals $[0, L]$, $[L, T]$ are small enough. These rather local results do not come close to covering the cases where invariant imbedding is applicable, and we shall not pursue this topic.

Example 3.1.1. Nonsmooth Characteristics

As a first application, let us consider use of invariant imbedding for differential equations with continuous but only piecewise differentiable solutions. Suppose, for example, the function F and G of (3.1.1) are differentiable in u and x and continuous in t except at some intermediate point $L \in (0, T)$. If u and x are required to be continuous at L, then Theorem 3.1.1 requires the solution of

$$u_{1t} + u_{1x_1} G_1(t, u_1, x_1) = F_1(t, u_1, x_1)$$
$$u_1(0, x_1) = f(x_1)$$

over $[0, L]$ and

$$u_{2t} + u_{2x} G_2(t, u_2, x_2) = F(t, u_2, x_2)$$
$$u_2(L, x_2) = u_1(L, x_2)$$

over $[L, T]$. In other words, we can solve the Cauchy problem

$$u_t + u_x G(t, u, x) = F(t, u, x), \qquad u(0, x) = f(x)$$

over $[0, T]$ provided we take into account that u_t is discontinuous at $t = L$. Thus, the invariant imbedding formulation of Section 2.2 remains unchanged for functions which are only piecewise continuous in t. We note that the numerical methods of Section 2.4 are not affected by a finite number of such discontinuities, as long as the points of discontinuity coincide with mesh points.

While discontinuities of F and G in t do not affect the invariant imbedding technique, discontinuities in u and x are quite a different matter. Even for a smooth initial manifold given parametrically by $\{u(0) = f(s), x(0) = s\}$ the characteristics, although continuous in t, will generally not be continuous in s and hence not generate any surface as s ranges of E^n. As an illustration, consider the scalar case

$$u' = 1, \qquad\qquad u(0) = 0$$

$$x' = \begin{cases} 1, & x > 0, \\ -1, & x \le 0, \end{cases} \quad x(0) = s$$

it is readily seen that as s ranges from $-\infty$ to $+\infty$ the functions $\{u(t, s),$ $x(t, s)\}$ do not define a surface in the sector $|x| < t$. Hence invariant imbedding is no longer applicable. A related situation occurs when F and G are continuously differentiable but the initial manifold is discontinuous. In this case the characteristics are continuously differentiable in the initial value, but the initial value is varied discontinuously. The result is the formation of shocks or, again, gaps in the integral surface. In his von Neumann lecture, Lax (1969) discusses in detail the system

$$du/dt = 0, \qquad dx/dt = u$$

and the related partial differential equation

$$u_t + u_x u = 0$$

when the initial manifold is a step function. If $u(0, x) = H(-x)$, where H is the usual Heaviside unit-step function, then shocks will occur because $u(t, x)$ can be multiply defined. If $u(0, x) = H(x)$, there will be a gap. It is shown that gaps and shocks can be handled numerically by taking into account the conservation laws applicable to the physical system and re- flected in the partial differential equations. In fact, the conservation scheme outlined in Section 2.4 fits into this context. We refer the reader to the paper of Lax and its references for a discussion of hyperbolic equations with shocks. For our purposes we can only state the negative conclusion that invariant imbedding cannot be routinely applied when either F, G, or f are discontinuous in u or x. ∎

Let us conclude this section with some comments on possible numerical difficulties. Clearly, the comments of Section 2.4 apply to the integration of the integral surfaces of (3.1.10) and (3.1.12) and, once $u_2(T, x_2)$ is found, to the solution of (3.1.13). The added complication for interface problems is due to Eqs. (3.1.11). As we have seen in Example 3.1.1, they do not come into play when the characteristics are continuous. On the other extreme, it may not be possible to find the expressions (3.1.11) in analytic form. This will complicate the numerical solution. Suppose, for example, that the interface condition is given as

$$u_2(L) = f_2(u_1(L)), \qquad x_2(L) = h_2(u_1(L))$$

where, because of incompatible dimensions or otherwise, the function h_2 is not invertible. In this case Eq. (3.1.10) can be integrated in the usual manner over a given grid of mesh points. For each grid point x_j on the interface $t = L$, we obtain a corresponding surface point $u_1(L, x_j)$. These two values determine a point on the initial manifold for $u_2(t, x)$, namely the point $(f_2(u_1(L, x_j)), h_2(u_1(L, x_j)))$. In general, the points $\{h_2(u_1(L, x_j))\}$ will not be evenly spaced on the x_2-axis. Hence, to continue the surface $u_2(t, x_2)$ over $[L, T]$ numerically either a variable mesh spacing must be used or the initial manifold must be determined through interpolation at the mesh points x_j before the same algorithm can be used. For more than one-dimensional base characteristics, either approach appears formidable. If initial value problems for u_2, x_2 can be integrated stably, it is probably advantageous to continue over $[L, T]$ by integrating the characteristic equations subject to $(f_2(u_1(L, x_j)), h_2(u_1(L, x_j)))$.

3.2. LINEAR INTERFACE PROBLEMS

The discussion of linear two-point boundary value problems already indicates that a substantial simplification of the equations may be expected if the characteristic equations and boundary conditions are linear. Since linear interface problems occur frequently in technical applications, we shall formulate the corresponding invariant imbedding equations in detail.

The problem is given as follows:

$$\begin{aligned} u_i' &= A_i(t)u_i + B_i(t)x_i + F_i(t), \\ x_i' &= C_i(t)u_i + D_i(t)x_i + G_i(t), \end{aligned} \qquad i = 1, 2 \qquad (3.2.1)$$

with boundary conditions

$$u_1(0) = f_1 x_1(0) + a_1$$
$$g(u_2(T), x_2(T)) = 0 \qquad (3.2.2)$$

and the interface condition at $L \in (0, T)$

$$H_1\begin{pmatrix} u_1(L) \\ x_1(L) \end{pmatrix} + H_2\begin{pmatrix} u_2(L) \\ x_2(L) \end{pmatrix} = c \qquad (3.2.3)$$

where u_i, x_i are m_i- and n_i-dimensional vectors such that $m_1 + n_1 = m_2 + n_2$ and where A_i, B_i, etc. are matrices of consistent dimensions. The boundary condition $g(u_2(T), x_2(T)) = 0$ is not necessarily linear so that, strictly speaking, the boundary value problem is nonlinear. However, as long as the boundary and interface conditions are nonlinear at one point only, the reduction of the invariant imbedding equations to ordinary differential equations applies. Above, we have allowed the endpoint condition to be nonlinear. The example below describes the modification necessary when the interface condition is nonlinear.

Because of the affine relationship (1.2.4) between u_i and x_i, we can rephrase Theorem 3.1.1 for (3.2.1)–(3.2.3) without further ado as:

THEOREM 3.2.1. Assume that

(i)   the initial value problem

$$U_1' = B_1(t) + A_1(t)U_1 - U_1 D_1(t) - U_1 C_1(t) U_1, \qquad U_1(0) = f_1$$
$$v_1' = [A_1(t) - U_1(t)C_1(t)]v_1 - U_1(t)G_1(t) + F_1(t), \qquad v_1(0) = a_1$$

has a solution over $[0, L]$;

(ii)   the interface condition

$$H_1\begin{pmatrix} U_1(L)x_1 + v_1(L) \\ x_1 \end{pmatrix} + H_2\begin{pmatrix} u_2 \\ x_2 \end{pmatrix} = c$$

has a solution

$$u_2 = f_2 x_2 + a_2, \qquad x_1 = k x_2 + b$$

(iii)   the initial value problem

$$U_2' = B_2(t) + A_2(t)U_2 - U_2 D_2(t) - U_2 C_2(t) U_2, \qquad U_2(L) = f_2$$
$$v_2' = [A_2(t) - U_2(t)C_2(t)]v_2 - U_2(t)G_2(t) + F_2(t), \qquad v_2(L) = a_2$$

has a solution over $[L, T]$; and finally,

(iv)   the equation

$$g(U_2(T)x + v_2(T), x) = 0$$

has a root $\hat{x}$. Then the characteristic $\{u_2(t), x_2(t)\}$ obtained by integrating

(3.2.1) for $i = 2$ subject to $u_2(T) = U_2(T)\hat{x} + v_2(T)$, $x_2(T) = \hat{x}$ is a solution of (3.2.1)–(3.2.3) over $[L, T]$, while the characteristic $\{u_1(t), x_1(t)\}$ through $(U_1(L)[kx_2(L) + b] + v_1(L), \ kx_2(L) + b)$ is its continuation over $[0, L]$.

Example 3.2.1. Heat Conduction in a Composite Slab with Radiation at the Interface

If only one interface is present, and if the boundary conditions at 0 and T are linear, then by integrating the invariant imbedding equations from the right and left toward the common interface treatment of the linearity can be postponed until the affine transformations are known at the interface. For example, consider again the one-dimensional heat conduction equation for the composite slab; but suppose that heat transfer at the interface $\xi = L$ occurs by radiation only. Such a problem may be described with the following equations

$$\partial^2 u/\partial \xi^2 - \partial u/\partial t = 0, \qquad \xi \in [0, 2], \quad t > 0, \qquad \xi \neq 2 \qquad (3.2.4)$$

$$u(\xi, 0) = u_0(\xi)$$

$$u(0, t) = a, \qquad t > 0$$

$$(\partial u/\partial \xi)(2, t) = 0, \qquad t > 0$$

with the radiation interface condition at $\xi = 1$ given by

$$(\partial u/\partial \xi)(1-, t) = \frac{\partial u}{\partial \xi}(1+, t) = u^4(1+, t) - u^4(1-, t)$$

It may be noted that these equations apply to a composite slab consisting of two identical sections with perfect insulation at the right end and specified temperature at the left end. It is straightforward to extend this problem to a slab with n (≥ 1) different sections and more general boundary and interface conditions. It is essential, however, that at most, one of these is nonlinear.

Discretizing this system by replacing $\partial u/\partial t$ with its backward difference and replacing $du/d\xi$ by x, we obtain the first-order system at time $t = n \Delta t$

$$u' = x, \qquad\qquad u(0) = a$$

$$x' = (1/\Delta t)[u - u_{n-1}(\xi)], \qquad x(2) = 0$$

where u_0 is given, and the two interface equations

$$x(1-) = x(1+)$$

$$x(1-) = u^4(1+) - u^4(1-)$$

Comparing with (3.2.1), we find that over $[0, 1]$ we have to solve the invariant imbedding equations

$$U' = 1 - (1/\varDelta t)U^2, \qquad\qquad U(0) = 0$$
$$v' = - (1/\varDelta t)U(\xi)(v - u_{n-1}(\xi)), \qquad v(0) = a$$

If we now were to substitute the representation $u(1-) = U(1)x(1-) + v(1)$ into the interface condition, as outlined in Theorem 3.1.1, we would have to use as interface condition the expression

$$u^4(1+) = -(U(1)x(1+) + v(1))^4 + x(1+)$$

We see that $u(1+, x)$ no longer is a linear initial manifold so that $u(\xi, x)$ cannot be found from ordinary differential equations over $1 \leq \xi \leq 2$.

Instead, let us generate the integral surface through $u'(2) = 0$. Base and surface characteristics are now reversed. The affine transformation over $1 \leq \xi \leq 2$ is written as

$$x(\xi) = R(\xi)u(\xi) + w(\xi)$$

where R and w satisfy the initial value problems

$$R' = (1/\varDelta t) - R^2, \qquad\qquad R(2) = 0$$
$$w' = -R(\xi)w - (1/\varDelta t)u_{n-1}(\xi), \qquad w(2) = 0$$

The interface conditions become

$$x(1) \equiv x(1-) = x(1+) = R(1)u(1+) + w(1)$$

and

$$x(1) = \left(\frac{x(1) - w(1)}{R(1)}\right)^4 - (U(1)x(1) + v(1))^4 \qquad (3.2.5)$$

This last equation must be solved for the unknown $x(1)$. Once $x(1)$ is known, the complete solution $u_n(\xi)$ on $[0, 1]$ and $[1, 2]$ can be found from the reverse sweep

$$x' = (1/\varDelta t)[U(\xi)x + v(\xi) - u_{n-1}(\xi)], \qquad x(1) \text{ as found from (3.2.5)}$$
$$u_n(\xi) = U(\xi)x(\xi) + v(\xi) \qquad\qquad\qquad \xi \in [0, 1]$$

and

$$u_n' = R(\xi)u_n + w(\xi), \qquad u_n(1) = (x(1) - w(1))/R(1), \quad \xi \in [1, 2]$$

We observe that U and R are given in closed form as

$$U(\xi) = (\Delta t)^{1/2} \tanh\{\xi/(\Delta t)^{1/2}\}, \qquad R(\xi) = 1/(\Delta t)^{1/2} \tanh\{(\xi - 2)/(\Delta t)^{1/2}\}$$

The functions were computed once and stored. The remaining differential equations were integrated with the trapezoidal rule, while the nonlinear interface equation (3.2.5) was solved with Newton's method. Similarly to the radiation problem discussed in Section 2.9, it may be shown that this iteration will converge for the initial guess $x(1) = 0$. Figure 3.2.1 shows the

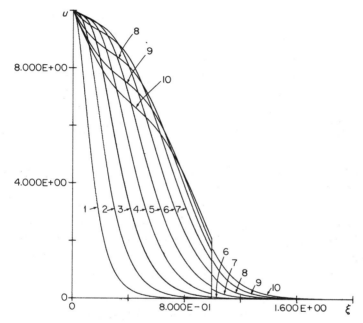

Fig. 3.2.1. Temperature in the composite slab for the first 10 time steps (curve number indicates time step).

computed temperature profile for the first 10 time steps when $u_0 = 0$, $a = 1$, and $\Delta t = 1$. Note that the initial/boundary conditions are discontinuous at $t = \xi = 0$. However, this caused no computational difficulties. Figure 3.2.2 traces the temperature after 20, 40, etc. time steps. ∎

In summary, the solution of the radiation problem indicates that invariant imbedding is well suited for interface problems when one of the interface or boundary conditions is nonlinear. All we need is the affine transformation at the nonlinear interface which is found by integrating the invariant imbedding equations from the right and left toward that particular interface. Once this

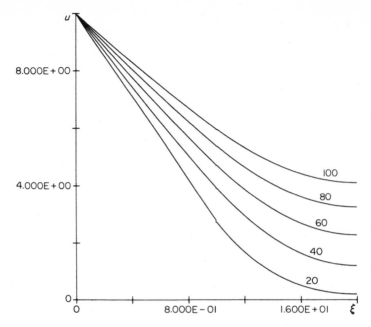

Fig. 3.2.2. Temperature in the composite slab for the first 100 times steps.

affine transformation is found, the nonlinear differential equations problem is reduced to a nonlinear n-dimensional algebraic problem. This approach is equivalent in a sense to solving the interface problem by finite differences and reducing the resulting algebraic equations by block Gaussian elimination such that the last block accounts for the nonlinearity.

3.3. FREE INTERFACE PROBLEMS

Many applications come to mind where the differential equations are subject to overspecified boundary and interface conditions, but where the location of the interface has to be determined as part of the problem. Perhaps the best known examples of this type of problem are the two-phase Stefan problems for heat transfer with change of phase. When the solution is known *a priori* on one side of the unknown interface, free interface problems will often reduce to free boundary problems. In that case, the methods of Section 2.10 would apply. For proper free interface problems the discussion has to be extended somewhat.

We shall consider the following systems

$$u_i{}' = F_i(t, u_i, x_i), \qquad x_i{}' = G_i(t, u_i, x_i) \tag{3.3.1}$$

with separated boundary conditions

$$u_1(0) = f_1(x_1(0)), \qquad u_2(T) = g_2(x_2(T)) \tag{3.3.2}$$

and with interface conditions

$$H(L, u_1(L), u_2(L), x_1(L), x_2(L)) = 0$$
$$h(L, u_1(L), u_2(L), x_1(L), x_2(L)) = 0 \tag{3.3.3}$$

Apart from the usual continuity assumptions on the given functions, we shall assume that (dimension u_1 + dimension x_1) = (dimension u_2 + dimension x_2) = n. Since we shall integrate from the right and left manifolds, it is not necessary that u_1 and u_2 are of identical dimensions. However, it is required that H defines a system of $n_1 + n_2$ (possibly nonlinear) equations, where n_i is the dimension of the base characteristic x_i, and that h defines a scalar equation. From a practical point of view it generally is preferable to single out the most complicated component of equation (3.3.3) as the scalar equation h because the numerical solution of a scalar equation on a bounded interval is little affected by its degree of nonlinearity if, for example, a bisection method is used. There are some variations possible in the description of the interface problem. In one application discussed in the following the function h is missing, instead an additional boundary value is given which is used to determine L. Such modifications of the basic problem (3.3.1)–(3.3.3) are, at least conceptually, easily handled provided there can exist only one interface (although its position need not be unique) and provided we consider invariant imbedding from the viewpoint of an implicit shooting method.

In order to obtain an algorithm for the solution of the free interface problem (3.3.1)–(3.3.3) we need only to rephrase Theorem 3.1.1.

THEOREM 3.3.1. Assume that

(i)   the initial value problems

$$u_{it} + u_{ix_i}(t, x_i)G(t, u_i, x_i) = F(t, u_i, x_i), \qquad i = 1, 2$$

subject to

$$u_1(0, x_1) = f_1(x_1), \qquad u_2(T, x_2) = g_2(x_2)$$

have solutions over $[0, T] \times E^{n_i}$, where n_i = dimension x_i, $i = 1, 2$;

(ii)   the $(n_1 + n_2 + 1)$-dimensional system

$$H(t, u_1(t, x_1), u_2(t, x_2), x_1, x_2) = 0$$
$$h(t, u_1(t, x_1), x_2(t, x_2), x_1, x_2) = 0$$

admits a solution $(L, \hat{x}_1, \hat{x}_2)$.

Then the solution of the initial value problem

$$u_i' = F_i(t, u_i, x_i), \qquad u_i(L) = u_i(L, \hat{x}_i)$$
$$x_i' = G_i(t, u_i, x_i), \qquad x_i(L) = \hat{x}_i$$

is a solution of the free interface problem (3.3.1).

Proof. The points found from (ii) satisfy the interface conditions (3.3.3). Since $u_1(L, \hat{x}_1)$ and $u_2(L, \hat{x}_2)$ also lie on the integral surfaces for (i), it follows that the characteristics through these points remain imbedded in $u_i(t, x_i)$ and hence pass through the given initial manifolds. By definition, the characteristics satisfy (3.3.1) and thus solve the free interface problem. █

It is possible to specify conditions for the data functions of (3.3.1)–(3.3.3) which ensure that each hypothesis of Theorem 3.3.1 can be satisfied. However, most of these conditions will be slanted toward a particular application. We shall not pursue these questions. Instead, several examples will be presented which illustrate the use of Theorem 3.3.1 for more concrete interface conditions.

Example 3.3.1. A One-Dimensional Scattering Problem

Suppose a particle passes through the origin at time $t = 0$ while moving under the influence of a known force depending on its speed and position. At an unknown time it collides with a known barrier and is scattered back with half its momentum. At a known future time T it passes again through the origin. We are interested in its initial momentum (speed). The corresponding free interface problem is

$$u'' = G(t, u, u')$$
$$u(0) = 0, \qquad u(T) = 0$$
$$u(L) = u_0, \qquad u'(L+) = -\tfrac{1}{2}u'(L-)$$

where u is the displacement, u' the speed of the particle, and u_0 the location of the barrier. If we reduce this system to the equivalent first-order system and identify u_1 and u_2 with the displacements before and after the collision, we obtain the following specific form for (3.3.1)–(3.3.3)

$$u_1' = x_1, \qquad u_1(0) = 0$$
$$x_1' = G(t, u_1, x_1)$$
$$u_2' = x_2, \qquad u_2(T) = 0$$
$$x_2' = G(t, u_2, x_2)$$

with interface condition

$$u_1(L) = u_2(L), \qquad x_2(L) = -\tfrac{1}{2}x_1(L)$$

According to Theorem 3.3.1, we have to solve

$$\partial u_1/\partial t + (\partial u_1/\partial x_1)G(t, u_1, x_1) = x_1, \qquad u(0, x_1) = 0$$
$$\partial u_2/\partial t + (\partial u_2/\partial x_2)G(t, u_2, x_2) = x_2, \qquad u_2(T, x_2) = 0$$

If we intersect to surfaces $u_i(t, x_i)$ with the plane $u = u_0$, we obtain two curves $x_i = x_i(t)$. The interface is now a root L of the equation

$$h(t) \equiv x_2(t) + \tfrac{1}{2}x_1(t) = 0$$

If such a point L exists, then the solution of

$$u' = x, \qquad u(L) = u_0$$
$$x' = F(t, u, x), \qquad x(L) = x_1(L)$$

integrated backward over $[0, T]$ yields the initial momentum $x(0)$. ∎

As we have observed many times before, the invariant imbedding equations are greatly simplified if the initial manifold and the characteristics equations are linear because in this case the representation

$$u_i = U_i(t)x_i + v_i(t) \tag{3.3.5}$$

can be substituted into the interface condition. If H is also linear, this substitution reduces the interface condition for determining the position of the interface to an explicit scalar equation in L, which, at least from a numerical point of view, is readily solved. Let us rephrase Theorem 3.3.1 for the linear problem

$$u_i' = A_i(t)u_i + B_i(t)x_i + F_i(t), \qquad u_i(b_i) = f_i x_i(b_i) + a_i$$
$$x_i' = C_i(t)u_i + D_i(t)x_i + G_i(t)$$

subject to the interface condition

$$H(t)\begin{pmatrix} u_1(t) \\ x_1(t) \\ u_2(t) \\ x_2(t) \end{pmatrix} = \phi(t)$$

$$h(t, u_1(t), x_1(t), u_2(t), x_2(t)) = 0$$

where H is of dimension $(n_1 + n_2) \times 2n$, where $\phi(t)$ is a given $(n_1 + n_2)$-dimensional vector defined on $[b_1, b_2]$ and where h is a continuous scalar function.

THEOREM 3.3.2. Assume that

(i)   the initial value problems

$$U_i' = B_i(b) + A_i(t)U_i - U_iD_i(t) - U_iC_i(t)U_i, \qquad U_i(b_i) = f_i$$
$$v_i' = [A_i(t) - U_i(t)C_i(t)]v_i - U_i(t)G_i(t) + F_i(t), \qquad v_i(b_i) = a_i$$

have solutions over $[b_1, b_2]$;

(ii)   the $(n_1 + n_2)$-dimensional system

$$H(t)\begin{pmatrix} U_1(t)x_1 + v_1(t) \\ x_1 \\ U_2(t)x_2 + v_2(t) \\ x_2 \end{pmatrix} = \phi(t)$$

has a solution $x_1 = \gamma_1(t)$, $x_2 = \gamma_2(t)$, and that the scalar equation

$$h(t, U_1(t)\gamma_1(t) + v_1(t), \gamma_1(t), U_2(t)\gamma_2(t) + v_2(t), \gamma_2(t)) = 0$$

has a solution $L \in (b_1, b_2)$. Then the boundary values $(u_i(L) = U_i(L)\gamma_i(L) + v_i(L), \gamma_i(L))$ reduce the free interface problem to two independent two-point boundary value problems.

For a proof we observe that this is precisely Theorem 3.3.1 when the affine transformation (3.3.5) for the integral surfaces is used. It may be noted that the only nonlinearity of the problem lies in solving the scalar equation $h(t) = 0$. In all examples below the invariant imbedding equations were integrated numerically, and the function $h(t) = 0$ was evaluated at each mesh point. When it changed sign, the root L was determined by linear interpolation. As a consequence, invariant imbedding lead to a numerical method for free interface problems which did not require any iteration. The following examples clarify the application of Theorem 3.3.2 to linear free interface problems.

Example 3.3.2. The Two-Phase Stefan Problem (See N.3.1)

Suppose a sphere is alternately heated and cooled on its surface. We shall assume that the system is described by the following equations:

$$\frac{\partial^2 u_i}{\partial r^2} + \frac{2}{r} \frac{\partial u_i}{\partial r} - \frac{\partial u_i}{\partial t} = f_i(t, r) \tag{3.3.6}$$

$$\frac{\partial u_1}{\partial r}(t, 0) = 0 \tag{3.3.7}$$

$$u_2(t, 1) = \alpha_2(t) \tag{3.3.8}$$

$$u_i(0, r) = u_0(r) \tag{3.3.9}$$

with the interface conditions

$$u_i(t, s(t)) = 0 \tag{3.3.10}$$

$$\frac{ds}{dt} + \frac{\partial u_1}{\partial r}(t, s(t)) - \frac{\partial u_2}{\partial r}(t, s(t)) = \mu(t, s(t)) \tag{3.3.11}$$

where u_1 and u_2 denote the temperature above and below melting, defined on $[0, s(t)]$ and $[s(t), 1]$, respectively, for $0 \le s(t) \le 1$. As in previous numerical schemes for the diffusion equation, the time derivative is replaced by a backward difference and the resulting second-order ordinary differential equations are written as a first-order system. For fixed n and known solutions $u_i^{n-1}(r)$ and s^{n-1}, we obtain on $[0, s]$ the expressions (after suppressing the superscript n)

$$x_1' = -(2/r)x_1 + (1/\Delta t)[u_1 - u_1^{n-1}(r)] + f_1(n \, \Delta t, r)$$
$$u_1' = x_1, \qquad x_1(0) = 0$$

Note that on $[0, s]$ the function $x_1(r) = du_1/dr$ is chosen as surface characteristic because its value is given at the origin. On the interval $[s, 1]$ we have the system

$$u_2' = x_2$$
$$x_2' = -(2/r)x_2 + (1/\Delta t)[u_2 - u_2^{n-1}(r)] + f_2(n \, \Delta t, r)$$
$$x_2(1) = \alpha(n \, \Delta t)$$

The interface condition (3.3.3) is of the form

$$u_1(s) = 0, \qquad u_2(s) = 0$$

$$F(s) \equiv s - s^{n-1} + \Delta t[x_1(s) - x_2(s)] - \Delta t \, \mu(n \, \Delta t, s) = 0$$

The surfaces $x_1(t, u_1)$ and $u_2(t, x_2)$ are given as

$$x_1(r, u_1) = U_1(r)u_1 + v_1(r)$$

$$u_2(r, x_2) = U_2(r)x_2 + v_2(r)$$

where

$$U_1' = 1/\Delta t - (2/r)U_1 - U_1^2, \qquad\qquad\qquad U_1(0) = 0$$

$$v_1' = -[U_1(r) + 2/r]v_1 - (1/\Delta t)u_1^{n-1}(r) + f_1(n \, \Delta t, r), \qquad v_1(0) = 0$$

$$U_2' = 1 + \frac{2}{r} U_2 - \frac{1}{\Delta t} U_2^2, \qquad\qquad\qquad U_2(1) = 0$$

$$v_2' = -\frac{1}{\Delta t}U_2(r)v_2 - \frac{1}{\Delta t}U_2(r)u_2^{n-1}(r) - U_2(r)f_2(n \, \Delta t, r), \qquad v_2(1) = \alpha(n \, \Delta t)$$

Substituting the representations for x_1 and u_2 into the interface condition $u_1(s) = u_2(s) = 0$ at an arbitrary point s leads to

$$x_1(s) = v_1(s), \qquad x_2(s) = -v_2(s)/U_2(s)$$

Substitution into the last interface equation requires that s be chosen such that

$$F(s) \equiv s - s^{n-1} + \Delta t[v_1(s) + v_2(s)/U_2(s)] - \Delta t \, \mu(n \, \Delta t, s) = 0$$

The root s^n of this equation is the interface at time $t = n \, \Delta t$, and the complete solution $u_1^n(r) \equiv u_1(r)$ and $u_2^n(r) \equiv u_2(r)$ at the nth time level is found by integrating

$$u_1' = U_1(r)u_1 + v_1(r), \qquad u_1(s^n) = 0$$

$$x_2' = [(1/\Delta t)U_2(r) - 2/r]x_2 + (1/\Delta t)[v_2(r) - u_2^{n-1}(r)] + f_2(n \, \Delta t, r)$$

$$x_2(s^n) = -v_2(s^n)/U_2(s^n)$$

and setting

$$u_2(r) = U_2(r)x_2(r) + v_2(r)$$

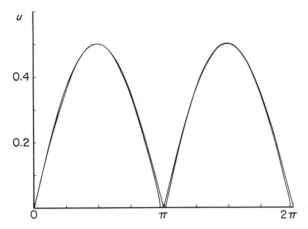

Fig. 3.3.1. Two-phase Stefan problem—periodic solution.[†]

Figures 3.3.1, 3.3.2, and 3.3.3 show the results of a numerical study when $f_i(t, r) = -(6 + \frac{1}{2} \sin t \cos t), \alpha(t) = \frac{1}{4} \sin^2 t - 1, \mu(t, r) = (\text{sign} \sin t) \frac{1}{2} \cos t$. The resulting differential equations were solved with the trapezoidal rule which has the advantage that the algebraic equations, being either linear or quadratic, can be solved in closed form. For the first illustration the initial conditions $u_2^0(r) = -r^2$, $s(0) = 0$ were chosen. In this case the Stefan problem has the periodic analytic solution $u_1(t, r) = u_2(t, r) = \frac{1}{4} \sin^2 t - r^2$ so that $s(t) = |\frac{1}{2} \sin t|$. Note that $s(t)$ is neither monotone nor everywhere

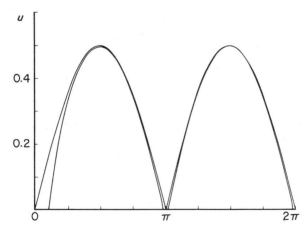

Fig. 3.3.2. Two-phase Stefan problem—liquid phase initially absent.

[†] Figs. 3.3.1–3.3.3 reprinted with permission from G. Meyer, *SIAM J. Numer. Anal.* **8**, 555–568 (1971). Copyright 1971 by Society for Industrial and Applied Mathematics.

differentiable. Figure 3.3.1 shows the computed and exact solutions when $\Delta r = 10^{-3}$ and $\Delta t = \pi/200$.

Figure 3.3.2 is obtained with the same parameters for an initial condition $u_2{}^0(r) = -2$ on $[0, 1]$. Note that in this case no liquid phase is present so that $(\partial u_2/\partial r)(t, 0) = 0$ until $u_2(t, 0) = 0$ and the second phase appears. Until this occurs we only have to solve the equations for u_2, x_2 subject to

$$x_2(0) = 0$$

It is seen from the illustration that the periodic solution is quickly established.

For Figure 3.3.3, the initial conditions $u_1(0, r) = 2$, $0 \leq r \leq \tfrac{1}{2}$, $u_2(0, r) = -2$, $\tfrac{1}{2} \leq r \leq 1$, $s(0) = \tfrac{1}{2}$ were chosen. Again the periodic solution predominated after approximately 100 time steps.

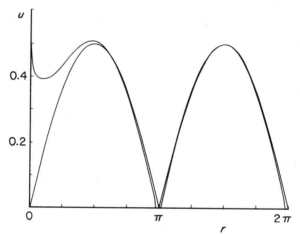

Fig. 3.3.3. Periodic Stefan problem—both phases initially present.

The deviation of the computed solution from the correct curve when $s(t) \approx 0$ is probably due to errors in integrating the Riccati equation near $r = 0$; the accuracy can likely be improved if instead of the numerical solution the analytic solution (see Table 2.7.1)

$$y = u(r) - 1/ar$$

of

$$y' + (2/r)y + b + ay^2 = 0$$

is chosen where

$$u' + b + au^2 = 0$$

Note (again from Table 2.7.1) that u subject to $u(r_0) = u_0$ has the solution

$$u(r) = \frac{u_0(-ab)^{1/2} + b \tan(-ab)^{1/2}(r - r_0)}{(-ab)^{1/2} + au_0 \tan(-ab)^{1/2}(r - r_0)}$$

The proper initial value u_0 at $r_0 = 0$ is determined from $u(0) = \lim_{r \to 0}$ $(y(r) + 1/ar)$. In this particular instance we have $y(0) = 0$ so that $u(0) = \infty$; here it will be advantageous to compute instead $(1/u)$ which satisfies the equation

$$(1/u)' + b(1/u^2) + a = 0, \qquad (1/u)(0) = 0.$$

Since $(1/u)$ is bounded away from 0 on $[\varepsilon, 1]$ for $\varepsilon > 0$, the function u is readily recovered.

Use of such analytic solutions can materially speed up numerical schemes for fixed and free boundary/interface problems. A more detailed discussion of this algorithm for one-dimensional two-phase and multiphase Stefan problems including existence, uniqueness, and convergence proofs may be found in Meyer (1971). Furthermore, similar problems arise when studying the precipitation products in a two-phase chemical system (Cannon and Hill, 1970), one-dimensional water flooding of oil reservoirs consisting of a compressible porous medium (Ciment and Guenther, 1969), the filtration of chemicals (Wentzel, 1960), fluidized bed coating (Elmas, 1970), viscoplastic impact (Kruzhov, 1967), or statistical decision theory (Sackett, 1971). All of these problems involve the solution of the one-dimensional diffusion equation subject to boundary conditions and an interface condition of the form (3.3.4) and (3.3.5). Invariant imbedding will apply to such problems. ∎

Example 3.3.3. A Shaft-Type Furnace (see N.3.2)

Consider a one-dimensional furnace shaft extending from $z = 0$ to $z = Z$. A cold solid is introduced at the top ($z = 0$) and heated by a hot gas rising from the bottom. At some intermediate point the descending solid reaches a threshold temperature above which it reacts with the gas and evolves heat. The physical system is indicated in Fig. 3.3.4. If steady-state conditions prevail, an energy balance for the system leads to the following equations for the temperatures T and θ of the gas and solid.

$$dT/dz + \alpha(T - \theta) = 0$$
$$d\theta/dz + \beta(\theta - T) = Q(\theta) \qquad (3.3.12)$$
$$T(Z) = T_z, \qquad \theta(0) = \theta_0$$

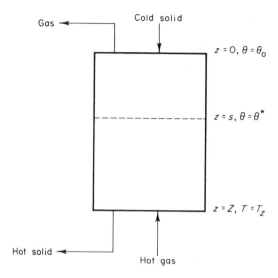

Fig. 3.3.4. Diagram of a shaft-type furnace.

where Q is a source term defined by

$$Q(\theta) = \begin{cases} 0, & \theta \leq \theta^* \\ Q_0, & \theta > \theta^* \end{cases}$$

Problem (3.3.12) is readily formulated as a free interface problem

$$T_i' + \alpha_i(T - \theta) = 0$$
$$\theta_i' + \beta_i(\theta - T) = F_i$$
$$F_1 = 0, \qquad F_2 = Q_0$$
$$\theta_1(0) = \theta_0, \qquad T_2(Z) = T_z$$

and interface condition at $z = s$

$$T_1(s) = T_2(s), \qquad \theta_1(s) = \theta_2(s) = \theta^*$$

Because $\theta_1(0)$ and $T_2(Z)$ are given, we choose the affine mappings

$$\theta_1 = U_1(z)T_1 + v_1, \qquad T_2 = U_2(z)\theta_2 + v_2$$

where U_i and v_i satisfy the initial value problems (2.2.2) and (2.2.3), in this case

$$U_1' = \beta_1 + (\alpha_1 - \beta_1)U - \alpha_1 U^2, \qquad U_1(0) = 0$$
$$v_1' = -(\beta_1 + \alpha_1 U_1(z))v_1, \qquad v_1(0) = \theta_0$$

and

$$U_2' = \alpha_2 - (\alpha_2 - \beta_2)U_2 - \beta_2 U_2^2, \qquad U_2(Z) = 0$$
$$v_2' = -[\alpha_2 + \beta_2 U_2(z)]v_2 - U_2(z)Q_0, \qquad v_2(Z) = T_z.$$

Note that the Riccati equations are well behaved. To find the interface s we have to determine a root of the system

$$\theta^* = U_1(s)T(s) + v_1(s)$$
$$T(s) = U_2(s)\theta^* + v_2(s)$$

Elimination of $T(s)$ leads to the following equation for s

$$F(s) \equiv [1 - U_1(s)U_2(s)]\theta^* - U_1(s)v_2(s) - v_1(s) = 0 \qquad (3.3.13)$$

We observe that $F(0) = \theta^* - \theta_0 > 0$ and $F(Z) = \theta^* - U_1(Z)T_z$. If $F(Z) \leq 0$, then this problem has at least one interface because $F(s)$ is defined and continuous on $[0, Z]$. Moreover, every root of $F(s) = 0$ defines an admissible interface. Unfortunately, lack of representative parameters for a furnace application precluded any sample calculations although we note that $F(Z)$ will be negative for sufficiently large T_Z because $U_1(Z) > 0$.

The system (3.3.12) is of particular interest because the transient analog of the shaft furnace problem leads to a hyperbolic free interface problem

$$a_i \frac{\partial T_i}{\partial t} + b_i \frac{\partial T_i}{\partial z} + c_i(T_i - \theta_i) = 0$$
$$(1 - d_i)e_i \frac{\partial \theta_i}{\partial t} + f_i \frac{\partial \theta_i}{\partial z} + g_i(\theta_i - T_i) = F_i$$

subject to appropriate initial, boundary, and interface conditions. It is readily verified that discretizing time and replacing $\partial/\partial t$ by a backward difference quotient will lead to a free interface problem like (3.3.12) at each time step. Its numerical solution by our algorithm should be rapid. However, no numerical calculations were carried out. ∎

Example 3.3.4. Overspecified Boundary Conditions

As a third problem, which at the same time points to an extension of the theory presented above, consider the motion of a particle through a two-layered liquid column. Suppose a particle begins its descent from position $u = 1$ at time $t = 0$. It falls under the influence of gravity while its motion is retarded through viscous drag which is proportional to its velocity. At time T, it reaches the origin $u = 0$. We intend to find the location of the interface between the lighter and heavier fluid in the column. The equations of motion are

$$u_i'' = -g - \alpha_i u_i'$$

$$u_1(0) = 1, \qquad u_1'(0) = 0, \qquad u_2(T) = 0$$

(3.3.14)

where g is gravity and where α_1, α_2 are viscosity dependent. At the (unknown) interface the motion is continuously differentiable, hence

$$u_1(L) = u_2(L), \qquad u_1'(L) = u_2'(L)$$

where L is the moment of passage through the interface.

These equations do not quite fit the expressions (3.3.1)–(3.3.3) because we have three boundary rather than three interface conditions. The initial conditions for u_1 determine completely the solution $u_1(t)$ for $t \geq 0$. However, this does not help in linking the solution with u_2 at the unknown interface. (The fact that u_1 is available in closed form is coincidental and will not be used in our analysis. Of course, for numerical work this may suggest a different solution technique for this particular example than invariant imbedding.) Moreover, for systems of higher dimensions not all initial conditions will generally be given so that the solution u_1 will not be available. On the other hand, the algorithm presented here can readily be extended to higher dimensional problems. The key to the solution is the transfer of one boundary condition to an interface condition which can be carried through as follows.

The solution u_2 has the representation

$$u_2(t) = R_2(t)u_2'(t) + v_2(t) \tag{3.3.15}$$

where as usual R_2 is the solution of the Riccati equation derived from the equivalent first-order system

$$u_i' = x_i, \qquad x_i' = -g - \alpha_i x_i$$

namely

$$R_2' = 1 + \alpha_2 R_2, \qquad R_2(T) = 0$$

while v_2 satisfies

$$v_2' = R_2(t)g, \qquad v_2(T) = 0$$

Similarly, for the region of liquid 1 we have *two* representations obtained by interchanging the definition of base and surface characteristics, namely

$$u_1(t) = R_1(t)u_1'(t) + v_1(t)$$

$$u_1'(t) = S_1(t)u_1(t) + w_1(t)$$

where

$$R_1' = 1 + \alpha_1 R_1, \qquad R_1(0) = 0$$

$$v_1' = R_1(t)g, \qquad v_1(0) = 1$$

and

$$S_1' = -\alpha_1 S_1 - S_1^2, \qquad S_1(0) = 0$$

$$w_1' = -[\alpha_1 + S_1]w_1 - g, \qquad w_1(0) = 0$$

The interface conditions now require that

$$R_1(L)u_2'(L) + v_1(L) = R_2(L)u_2'(L) + v_2(L)$$

$$u_2'(L) = S_1(L)u_2(L) + w_1(L)$$

$$= S_1(L)R_2(L)u_2'(L) + S_1(L)v_2(L) + w_1(L)$$

Eliminating $u_2'(L)$ from these two equations, we see that L has to be chosen as a root of the scalar equation

$$F(t) \equiv (R_1(t) - R_2(t))(S_1(t)v_2(t) + w_1(t))$$

$$- (1 - S_1(t)R_2(t))(v_2(t) - v_1(t)) = 0$$

We have presented these equations in detail to show that their derivation does not depend in any way on the structure of (3.3.14) other than its linearity. For this particular problem all of the equations have closed form solutions. In particular, we see by inspection that $S_1 \equiv 0$ and that $F(0) = 1 - v_2(0)$, $F(T) = R_1(T)w_1(T) + v_1(T)$. Furthermore, it is straightforward to verify that

$$R_1(t) = (e^{\alpha_1 t} - 1)/\alpha_1$$

$$v_1(t) = 1 + [(1/\alpha_1^2)(e^{\alpha_1 t} - 1) - (1/\alpha_1)t]g$$

$$R_2(t) = (e^{\alpha_2(t-T)} - 1)/\alpha_2$$

$$v_2(t) = [(e^{\alpha_2(t-T)} - 1)/\alpha_2^2 - (t - T)/\alpha_2]g,$$

$$w_1(t) = -g[1 - \exp(-\alpha_1 t)]/\alpha_1$$

From these expressions we obtain

$$F(0) = 1 - g[(e^{-\alpha_2 T} - 1)/\alpha_2{}^2 + T/\alpha_2)$$

$$F(T) = 1 + (g/\alpha_1{}^2)[1 - \alpha_1 T - \exp(-\alpha_1 T)]$$

We note that $F(0) > 0$ as $\alpha_2 \to \infty$ and $F(T) \to 1 - gT^2/2$ as $\alpha_1 \to 0$. Hence we can always assure a solution of $F(t) = 0$ provided the lower fluid is sufficiently thick and the upper fluid sufficiently thin as long as the time of passage T exceeds that required for a free fall over a unit distance under the action of gravity g. This is well in accord with physical intuition. Conversely, if for given α_1, α_2 the time span T is too large, then $F(0) < 0$ and $F(T) < 0$ and (3.3.14) need not have a solution. This, too, is in accord with intuition. ∎

As an aside, it may be mentioned here that the transfer of initial conditions to boundary or interface conditions may also be used, at least in principle, for the integration of pure initial value problems. As an illustration, consider the scalar system

$$u' = F(t, u, x), \qquad u(0) = a$$

$$x' = G(t, u, x), \qquad x(0) = b$$

Considering first u and then x as surface characteristic, we find that the solution is imbedded in the integral surfaces $u(t, x)$ and $x(t, u)$ of

$$\partial u/\partial t + (\partial u/\partial x)G = F, \qquad u(0, x) = a$$

and

$$\partial x/\partial t + (\partial x/\partial u)F = G, \qquad x(0, u) = b$$

Suppose now that for arbitrary T the system

$$u = u(T, x), \qquad x = x(T, u)$$

has a solution $\hat{u}$, $\hat{x}$, then the characteristic through $(\hat{u}, \hat{x})$ is a solution of the initial value problem. However, not only does this approach appear to be rather circuitous, but also suspect from a stability point of view. For example, for the system

$$u' = x, \qquad x' = u$$

we have the representation $u = U(t)x + v(t)$, $x = R(t)u + w(t)$ so that at an arbitrary point t

$$u = (U(t)w(t) + v(t))/(1 - U(t)R(t))$$
$$x = (R(t)v(t) + w(t))/(1 - R(t)U(t))$$

Since $U(t) = R(t) = \tanh t$, if follows that $1 - U(t)R(t) \rightarrow 0$ as t becomes large. Of course, if a free boundary is to be determined, this approach is useful, because we can multiply through by $1 - U(t)R(t)$. For example, if we are to find the boundary T where $u(T) = \alpha$, we have to solve the equation

$$F(t) \equiv (1 - U(t)R(t)\alpha - (U(t)w(t) + v(t)) = 0$$

In contrast to a search based on the known initial value solution for $u(t)$ and $x(t)$, no exponentially growing solutions occur in the scalar function F. For further comments on the application of invariant imbedding to (possibly unstable) initial value problems we refer to Scott (1970).

3.4. DYNAMIC PROGRAMMING

In Section 3.1, it was observed that the invariant imbedding equations remain valid when the characteristic equations are continuous but only piecewise differentiable. This observation can be used to derive the dynamic programming equation of Bellman for certain continuous optimization problems. It should be emphasized, however, that this section is intended only to establish the natural connection between invariant imbedding and its generalization, dynamic programming. There exist many texts on the theory and application of dynamic programming [see, e.g., Bellman (1957) or Lee (1968)] which the reader interested in optimization via dynamic programming may wish to consult.

To show how the dynamic programming equation can be derived from the invariant imbedding formulation, let us consider the general control problem

$$J' = F(t, J, x, u), \qquad J(T) = g(x(T))$$
$$x' = G(t, J, x, u), \qquad x(0) = f(J(0)) \tag{3.4.1}$$

where J is the cost functional, x the n-dimensional state vector and where u belongs to a specified set K of admissible control functions defined on $[0, T]$. We shall assume that F and G are continuous in all argument and continuously differentiable in J and x. The objective is to minimize $J(0)$ over the set of controls K. We shall require here that K is the set of piecewise continuous functions on $[0, T]$. This is a stringent condition and will be relaxed for certain problems in Section 5.3.

For a fixed control $u \in K$ we see that Eq. (3.4.1) is a two-point boundary value problem. As indicated by Example 3.1.1, its solution is imbedded into the piecewise differentiable integral surface $J(t, x, u)$ of

$$\frac{\partial J}{\partial t}(t, x, u) + J_x(t, x, u)G(t, J, x, u) = F(t, J, x, u)$$

$$J(T, x) = g(x) \tag{3.4.2}$$

and has to satisfy the equation

$$x = f(J(0, x, u(0))) \tag{3.4.3}$$

If the set of solutions of (3.4.3) is denoted by $S(u)$, then the optimum value $\hat{J}(0)$ for the given control u can be characterized as

$$\hat{J}(0, u) = \inf\{J(0, x, u(0)): x \in S(u)\}$$

In many applications the state vector is subject to a given initial value $x(0) = x_0$. In this case, of course, $S(u) = x_0$ and $\hat{J}(0, u) = J(0, x_0, u)$.

Proceeding one step further, we see that the optimum $\hat{J}(0, \hat{u})$ over all admissible controls can be characterized as follows:

$$\hat{J}(0, \hat{u}) = \inf\{\hat{J}(0, u): u \in K\} = \inf_{u \in K}\{\inf_{x \in S(u)} J(0, x, u)\} \tag{3.4.4}$$

Again, if $x(0) = x_0$ this condition reduces to the expression

$$\hat{J}(0, \hat{u}) = \inf\{J(0, x_0, u): u \in K\}.$$

For reasons to be given, Eq. (3.4.4) may be called a generalized dynamic programming equation.

Let us suppose that $x(0) = x_0$. Then (3.4.4) states that the optimum control $\hat{u}$ yields that integral surface $J(t, x, \hat{u})$ which assumes its minimum at $t = 0$, $x = x_0$, and $u(0) = \hat{u}(0)$. Consider now the cross section through $J(t, x, u)$ along $x = x_0$ for arbitrary $u \in K$. If for given $u_1, u_2 \in K$ there can be synthesized a third control function $u_3 \in K$ such that

$$J(t, x_0, u_3) \leq \min\{J(t, x_0, u_1), J(t, x_0, u_2)\}$$

then a sufficient condition for $\hat{u}$ to be optimal is that $J(t, x_0, \hat{u})$ is the lower envelope to the family $\{J(t, x_0, u): u \in K\}$. This lower envelope can be obtained by finding the control function $\hat{u}$ for which

$$J_t(t, x_0, \hat{u}) \geq J_t(t, x_0, u)$$

for all $u \in K$. In other words, $\hat{u}$ is a solution of the dynamic programming equation of Bellman (1957, p. 264).

$$J_t(t, x_0, \hat{u}) = \sup_{u \in K}\{F(t, J, x_0, u) - J_x(t, x_0, u)G(t, J, x_0, u)\}$$

$$J(T, x) = g(x)$$

(3.4.5)

where J_t is continuous at the points of continuity of u. Equation (3.4.5) only makes sense when any two controls can be synthesized to form a better approximation to the lower envelope. On the other hand, Eq. (3.4.4) still characterizes the optimal control even when no synthesis is possible as in the case of only finitely many controls in K. Of course, both (3.4.4) and (3.4.5) represent sufficient conditions for optimality.

Example 3.4.1. Linear State and Cost Equations

As an application, let us consider a linear control problem which we shall write as

$$J(0) = \langle y, x(T) \rangle + \int_0^T [a(t)J + \langle b(t), x(t) \rangle + F(t, u)] \, dt$$

$$x' = c(t)J + D(t)x + G(t, u), \qquad x(0) = x_0$$

where y, $b(t)$, and $c(t)$ are given continuous n-dimensional column vectors, where D is a continuous $n \times n$ matrix, and where F and G are assumed to be piecewise continuous for all $u \in K$. We again intend to minimize $J(0)$ over K. Replacing the lower limit of integration by t we see that this problem is equivalent to

$$J' = -a(t)J - \langle b(t), x \rangle - F(t, u), \qquad J(T) = \langle y, x(T) \rangle$$

$$x' = c(t)J + D(t)x + G(t, u), \qquad x(0) = x_0$$

This is a linear two-point boundary value problem and for arbitrary u admits the representation

$$J(t, x, u) = \langle U(t)^{\mathrm{T}}, x(t) \rangle + v(t, u)$$

(3.4.6)

where the matrix $U = (U_1, \ldots, U_n)$ satisfies the $(1 \times n)$-dimensional Riccati equation (2.6.2), in this case

$$U' = -b(t)^{\mathrm{T}} - a(t)U - UD(t) - Uc(t)U, \qquad U(T) = y^{\mathrm{T}}$$

while v is found from the scalar equation

$$v' = -[U(t)c(t) + a(t)]v - F(t, u) - U(t)G(t, u), \qquad v(T) = 0$$

We see that U is not influenced by the choice of control and can be computed immediately. Once U is known the linear equation for v has the solution

$$v(t) = \int_t^T \phi(T, r)[F(r, u) + U(r)G(r, u)]\, dr$$

where $\phi(t, r)$ is the positive fundamental solution of $v' = -[U(t)c(t) + a(t)]v$ $v(0) = 1$.

The generalized dynamic programming equation requires that $\hat{u}$ be chosen such that

$$v(0, \hat{u}) = \inf v(0, u)$$

$$= \inf \int_0^T \phi(T, r)[F(r, u) + U(r)G(r, u)]\, dr \qquad (3.4.7)$$

We note here that F and G only need be integrable in order for $v(0, u)$ to be defined. This will allow us to consider more general controls than only piecewise differentiable functions. We shall return to this point in Chapter 5. Finally, we note that Bellman's equation for (3.4.6), i.e., Eq. (3.4.7), only involves the function v. In fact, it follows from (3.4.7) that $\hat{u}$ has to be chosen such that the functional $f: K \rightarrow R$ satisfies

$$f(\hat{u}) = \inf_{u \in K} [F(t, u(t)) + U(t)G(t, \hat{u}))], \qquad t \in [0, T]$$

The minimization of a functional over a specified class of admissible functions is a well studied subject and for a suitable structure of F and G one can obtain existence theorems for the occurrence of $\hat{u}$. We shall not pursue this problem or the problem of actually finding $\hat{u}$ analytically or numerically (see N.4.1). ▮

Example 3.4.2. Linear States and Quadratic Cost

A similar approach may be taken in the case of linear state equations and a quadratic cost function. Specifically, let us consider a case similar to the minimum fuel problem of Section 2.10, namely the linear regulator problem

$$J = \langle Sx(T), x(T) \rangle + \int_0^T [\langle Q(t)x, x \rangle + \langle R(t)u, u \rangle] \, dt$$

$$x' = A(t)x + B(t)u, \qquad x(0) = x_0$$

(3.4.8)

where T is a given final value. Again, we intend to minimize J over the set K of piecewise continuous functions defined on $[0, T]$.

If we define $J(t)$ as

$$J(t) = \langle Sx(T), x(T) \rangle + \int_t^T [\langle Q(s)x, x \rangle + \langle R(s)u, u \rangle] \, ds$$

then for given $u \in K$ the functions J and x have to satisfy the two-point boundary value problem

$$J' = -\langle Q(t)x, x \rangle - \langle R(t)u, u \rangle, \qquad J(T) = \langle Sx(T), x(T) \rangle$$

$$x' = A(t)x + B(t)u, \qquad\qquad x(0) = x_0$$

Invariant imbedding assures that $J(0) = J(0, x_0, u)$ can be found from the Cauchy problem

$$\partial J/\partial t + J_x[A(t)x + B(t)u] = -\langle Q(t)x, x) \rangle - \langle R(t)u, u \rangle$$

$$J(T, x, u) = \langle Sx, x \rangle$$

The solution of this initial value problem is unique, at least provided T is sufficiently small. Moreover, it follows by inspection that J can be assumed to have the representation

$$J(t, x, u) = \langle U(t)x, x \rangle + \langle v(t), x \rangle + z(t)$$

(3.4.9)

where

$$U' = -Q(t) - UA(t) - A^{\mathrm{T}}(t)U, \qquad U(T) = S$$

$$v' = -A^{\mathrm{T}}(t)v - [U(t)B(t) + B^{\mathrm{T}}(t)U(t)]u(t), \qquad v(T) = 0$$

$$z' = -\langle v(t), B(t)u \rangle - \langle R(t)u, u \rangle, \qquad z(T) = 0$$

For given $u \in K$, the functions $U(t)$, $v(t)$, and $z(t)$ are well defined, and it is readily verified by backsubstitution and use of the defining equations for U, v, z that (3.4.9) is indeed the unique solution of the invariant imbedding equation. We note that only v and z are functions of u so that the dynamic programming equation reduces to

$$f(\hat{u}) = \sup_{u \in K} \{\langle v(t, u), x_0 \rangle' + z'(t, u)\}$$

However, it is not yet clear how this equation is related to the usual dynamic programming optimization equations for the linear regulator as discussed in Bellman (1967, Chapter 8). ∎

3.5. OVERCOMING CRITICAL LENGTHS

As has been pointed out in Section 2.4, invariant imbedding requires special care if the interval of integration exceeds the critical length of the problem in question. For linear equations the critical length is reflected in unbounded solutions for the Riccati equation. For square systems, however, unboundedness of the solution usually entails boundedness of its inverse. This inverse satisfies the invariant imbedding equation when the definition of surface and base characteristic is reversed. Hence it follows that by redefining the surface characteristic before reaching the critical length, the inverse of the affine mapping (1.2.4) can be continued past the critical, length, provided, of course, that the inverse exists in a neighborhood of the critical point. The following numerical example shows that this qualification is necessary for multidimensional problems.

To be specific consider the n-dimensional problem (see N.5.1)

$$x'' + C(t)x = F(t), \qquad x(0) = 0, \qquad x(T) = 0 \tag{3.5.1}$$

where $C(t)$ is a positive definite matrix satisfying $m\| x \|^2 \leq \langle C(t)x, x \rangle \leq M\| x \|^2$ for $M \geq m > 0$ and all $t \in E^n$. We know from the general theory that $x(t)$ and $x'(t)$ are related through the transformation

$$x(t) = U(t)x'(t) + v(t)$$

where

$$U'(t) = 1 + UC(t)U, \qquad\qquad U(0) = 0$$
$$v' = U(t)C(t)v - U(t)F(t), \qquad v(0) = 0 \tag{3.5.2}$$

The second equation is linear and has a solution whenever $U(t)$ exists. We shall disregard it here.

Corollary 1.3.3 shows that $U(t)$ is Hermitian and bounded above and below by

$$(M^{-1/2} \tan m^{1/2}t) \| x \|^2 \leq \langle U(t)x, x \rangle \leq m^{-1/2} \tan M^{1/2}t$$

Clearly, if $m^{1/2}T \geq \pi/2$, the solution $U(t)$ does not exist on $[0, T]$. Let us place an interface at some $L_1 < \pi/2M^{1/2}$ and require that the solution $x(t)$ be continuously differentiable at L. Over $[L_1, T]$ we shall choose $x'(t)$ as the surface characteristic. Now the representation

$$x'(t) = W(t)x(t) + z(t)$$

holds over $[L_1, T]$, where

$$W' = -C(t) - W^2, \qquad W(L) = U^{-1}(L)$$
$$z' = -W(t)z + F(t), \qquad z(L) = -W(L)v(L)$$

(3.5.3)

Applying Corollary 1.3.3, we see that W will exist and remain bounded below at least on $[L, 2L]$. We continue $W(t)$ until it has a bounded inverse (preferably until it is negative definite). We then invert again and compute $U(t)$. In general, it follows that U and W exist and are computable on $[0, \infty)$ provided not both are singular at the same point. If T is a point where U is singular, the equation

$$x(T) = U(t)x'(T) + v(T)$$

may not be solvable or have multiple solutions. In particular, with given $x(T)$ there will exist multiple solutions if $U(T)$ is singular and $\langle y, x(T) - v(T) \rangle = 0$ for all y belonging to the null space $v(U^*(T))$ of $U^*(T)$. At all other points where $U(T)$ is nonsingular the boundary value problem (3.5.1) has a unique solution.

For scalar equations the points of singularity are readily obtainable because the solution of the Riccati equation passes through 0. This observation is the basis of the algorithm for the computation of eigenvalues for Sturm-Liouville problems developed by Scott *et al.* (1969). In the matrix case detection of points where the solution of the Riccati equation has a singular inverse may not be as straightforward. For small systems, however, many efficient algorithms exist for finding the spectrum of a given matrix, hence it is computationally quite feasible to find the points where the matrix is singular.

Example 3.5.1. A Three-Dimensional Eigenvalue Problem

As a illustration of the method of invariant imbedding in the presence of singular points let us determine the smallest eigenvalue λ of the linear problem

$$x'' + \lambda C(t)x = 0, \qquad x(0) = x(1) = 0 \qquad (3.5.4)$$

where

$$C(t) = \begin{pmatrix} 1 & \sin t & 0 \\ \sin t & 1.5 - e^{-t} & -\cos t \\ 0 & -\cos t & 1 \end{pmatrix}$$

The matrix C was chosen at random and does not describe any particular physical system. We shall attack this problem with a trial and error method by guessing λ, and two interfaces L_1 and L_2, computing $U(t)$ from (3.5.2) (with $C(t)$ replaced by $\lambda C(t)$) over $[0, L_1]$, $W(t)$ from (3.5.3) over $[L_1, L_2]$, and $U(t)$ from (3.5.2) subject to $U(L_2) = W^{-1}(L_2)$ over $[(L_2, 1]$. We then compute the spectrum of $U(1)$ and, if necessary, adjust λ and recompute $U(1)$ until at least one of the eigenvalues of $U(1)$ is zero. The choice of L_1 and L_2 depends on the critical length of (3.4.2) and is adjusted on the basis of the computed solution. Finally, we note that the eigenvalues of $U(1)$ grow monotonically with γ which simplifies the trial and error method.

Table 3.5.1

SPECTRUM OF THE RICCATI SOLUTION $U(1)$ AS A FUNCTION OF λ

λ	Eigenvalues of $U(1)$		
	μ_1	μ_2	μ_3
5.10	-0.5232	1.0620	-0.0012
5.11	-0.5203	1.0621	-0.0003
5.12	-0.5174	1.0623	0.0007
5.13	-0.5146	1.0625	0.0016

In our computation the Riccati (3.5.2) was integrated with a fourth order Runge–Kutta routine using a step size of $\Delta t = 10^{-2}$. For L_1 and L_2 we chose $t = 0.3$ and $t = 0.9$. Table 3.5.1 shows the spectrum of $U(1)$ as a function of λ. Starting at $t = 0$ the eigenvalues of $U(t)$ increased until one of them became infinite. Inversion prior to blowing up yielded three positive eigenvalues for $W(L_1)$. Continued integration forced two of the eigenvalues to become negative before $W(L_2)$ was inverted, yielding two negative and one positive eigenvalue for $U(L_2)$. As the integration continued, the eigenvalues increased again. It followed by interpolation between the values of μ_3 in Table 3.5.1 that a value of $\gamma \approx 5.1128$ furnishes a matrix $U(1)$ which

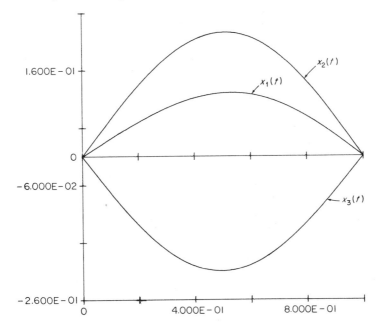

Fig. 3.5.1. Eigenvectors corresponding to the eigenvalue $\lambda = 5.1128$.

is very close to being singular. In fact, the corresponding μ_3 was computed to be

$$\mu_3 = 3.70 \times 10^{-6}$$

For the value of $\lambda = 5.1128$ the eigenvector $x'(1)$ corresponding to the computed eigenvalue μ_3 was determined. This vector was computed to be (after normalizing) (0.3783, 0.6968, −0.6094) and, together with $x(1) = 0$, provided the six initial values for a backintegration of (3.5.4). Figure 3.5.1 shows the curves $x(t)$ obtained with a Runge–Kutta method for $\Delta t = 10^{-2}$. The maximum component of $x(t)$ at $t = 0$ was 8×10^{-6}, while the exact solution of (3.5.4) should have been 0.

It is conceivable that the matrix $U(1)$ corresponding to a particular choice of $C(t)$ can have two distinct eigenvalues μ_1, μ_2 such that $\mu_1 \to 0$ and $\mu_2 \to \infty$ as $t \to 1$. In this case invariant imbedding would appear to be inapplicable.

While no eigenvalue below $\lambda = 5.1128$ exists, it is straightforward to find the next larger eigenvalue. Increasing λ will cause all three eigenvalues of $U(1)$ to grow. Hence λ has to be adjusted such that $\mu_1 \approx 0$. Since μ_2 never became negative on [0, 1], the next larger eigenvalues of (3.5.1) require additional interfaces. ∎

The above comments for linear systems apply equally well to certain non-linear equations. Consider the system

$$u' = F(t, u, x), \qquad\qquad u(0) = f(x(0))$$
$$x' = G(t, u, x), \qquad g(u(T), x(T)) = 0$$

where, besides the usual smoothness requirements, existence of unique characteristics through any point on the initial manifold will be postulated. Suppose that $u(T, x)$ exists for all $x \in E^n$, then the equation

$$H(u, x) \equiv u - u(T, x) = 0$$

defines implicitly an initial manifold when x is taken as surface characteristic. If we set $u_0 = u(T, x_0)$ for an arbitrary vector $x_0 \in E^n$, we can find this initial manifold explicitly from the implicit function theorem (Dieudonné, 1969, p. 270) as the solution of the differential equation

$$dx/du = u_x(T, x)^{-1}, \qquad x(u_0) = x_0$$

If u_x is nonsingular and Lipschitz continuous in x, then this equation can be integrated, at least in principle.

For general nonlinear systems this method is of doubtful computational value. It may, however, have its use for extending the existence theorems of Section 2.3. For example, the proof of Theorem 2.3.3 furnishes a uniform bound on u_x^{-1} which can be used to estimate the interval of existence for the integral surface $x(t, u)$ of

$$\partial x/\partial t + (\partial x/\partial u)F(t, u, x) = G(t, u, x)$$

through the computed initial manifold at $t = T$.

NOTES

N.1.1. In fact, for equations of divergence form the interface condition does not explicitly appear provided the interface location coincides with a mesh point in piecewise polynomial approximations [see Ciarlet *et al.* (1967)].

N.3.1. Because of their technical importance and mathematically interesting structure, two- and multiphase Stefan problems have been analyzed extensively from both the analytical and numerical point of view. For a classical approach to the Stefan problem we refer to Cannon *et al.* (1967);

a discussion of weak solutions may be found in Friedman (1968) and, for a related problem, in Cannon and Hill (1970). The numerical approach presented here is analyzed in detail in Meyer (1971) where additional references for the numerical solution of one-dimensional Stefan problems may be found. An application of this algorithm in a computer graphics parameter study for determining the depth of thaw under road surfaces on layered permafrost is discussed in Meyer *et al.* (1972).

N.3.2. The author is indebted to J. R. Cannon for bringing this example to his attention.

N.4.1. The derivation of the dynamic programming equation for linear states and costs is worked out in detail in Meyer (1970b) for Banach space valued evolution equations. In particular it is shown that the maximum principles of Friedman (1967) for fixed time control problems are recovered as special cases of the general theory.

N.5.1. This is certainly not the most general problem to which the technique of this section will apply. In fact, any system of equal surface and base characteristic dimension can be attacked from this point of view. The restriction to Hermitian systems allows us to draw on the existence theory of Section 1.3.

Chapter

4

Multipoint Boundary Value Problems

4.1. THE INVARIANT IMBEDDING EQUATIONS

Just like interface problems, multipoint boundary value problems for ordinary differential equations arise naturally in technical applications. For example, given a dynamical system with n degrees of freedom, there may be available exactly n states observed at n different times. A mathematical description of such a system results in an n-point boundary value problem. Another source of multipoint problems is the discretization of certain boundary value problems for partial differential equations over irregular domains with the method of lines. Strictly speaking, multipoint problems for ordinary differential equations are a particular class of interface problems and hence solvable with the techniques of the preceding chapter. There are, however, some conceptual differences, as well as notational complications which justify a separate treatment of multipoint problems.

If invariant imbedding is to be applied to multipoint boundary value problems, it may become necessary to redefine the surface and base characteristics at each data point which renders inadequate the notation used before for two-point boundary value problems, where the surface and base characteristics remained unchanged over the interval of integration. Here we shall work with a single n-dimensional system of ordinary differential equations

$$u' = F(t, u) \tag{4.1.1}$$

which in general may be subject to m vector boundary conditions of the form

$$f_i(u(t_i)) = 0 \qquad (4.1.2)$$

where f_i is a k_i-dimensional function such that $\sum_{i=1}^{m} k_i = n$. For ease of exposition, we shall assume that $m = n$ and $k_i = 1$. In other words, n scalar boundary equations of the general form (4.1.2) are given at n boundary points $\{t_i\}_{i=1}^{n}$. We shall agree that $t_i \in [t_1, t_n]$ for $i = 2, \ldots, n - 1$. If F is continuous in t and Lipschitz continuous in u then a solution of (4.1.1) and (4.1.2) is a continuously differentiable vector $u(t)$ that satisfies the differential equation for all $t \in [t_1, t_n]$ and the boundary conditions at t_i, $1 \leq i \leq n$.

If F is only piecewise continuous in t, then u is required to be differentiable at the points of continuity and continuous at the discontinuities.

It will be helpful to introduce a splitting of E^n into the direct sum $E^n = E_1^k \oplus E_2^k$, where E_1^k and E_2^k are k and $(n - k)$-dimensional Euclidean spaces. This splitting induces a splitting for elements of E^n; for arbitrary $u \in E^n$ we shall write $u = (u_1, \ldots, u_n)^T = u_1^k + u_2^k$, where $u_1^k = (u_1, \ldots, u_k)^T$ and $u_2^k = (u_{k+1}, \ldots, u_n)^T$. To provide a certain continuity in notation when referring to earlier sections, we shall make the identification

$$u_2^k \equiv x_2^k$$

In this notation then E_1^k is the space of the surface characteristic and E_2^k is the space of the base characteristic. Note that the spaces E_1^0 and E_2^n do not make sense; elements formally belonging to these spaces are to be ignored in the formulas given below.

Let us now derive the invariant imbedding equations for the multipoint boundary value problem (4.1.1) and (4.1.2). Suppose that the boundary value (4.1.2) for $i = 1$ can be solved for u_1 such that

$$u_1(t_1) = g_1(u_2(t_1), \ldots, u_n(t_1))$$

Using the shooting method we integrate (4.1.1) subject to

$$u_1^1(t_1) = g_1^1(s_2^1)$$
$$u_2^1(t_1) \equiv x_2^1(t_1) = s_2^1$$

Let u_1^1 and x_2^1 be the one-dimensional surface and $(n - 1)$-dimensional base characteristics. It follows from Theorem 1.1.2 that the solution of

(4.1.1) and (4.1.2) is imbedded into the integral surface $u_1^{1}(t, x_2^{1})$ of

$$\partial u_1^{1}/\partial t + (\partial u_1^{1}/\partial x_2^{1})F_2^{1}(t, u_1^{1}, x_2^{1}) = F_1^{1}(t, u_1^{1}, x_2^{1})$$

$$u_1^{1}(t_1, x_2^{1}) = g_1(x_2^{1})$$

where F_1^{1} and F_2^{1} are the components of F in E_1^{k} and E_2^{k}, respectively. Suppose that this integral surface exists over $[t_1, t_2]$ for all $x_2^{1} \in E_2^{1}$. At $t = t_2$, it is seen from (4.1.2) that the component $u_2(t_2)$ has to be chosen such that

$$f_2(u_1^{1}(t_2, u_2, x_2^{2}), u_2, x_2^{2}) = 0 \qquad (4.1.3)$$

We shall postulate the existence of a solution

$$u_2(t_2) = g_2(x_2^{2})$$

for Eq. (4.1.3). If we let u_1^{2} and x_2^{2} be the two-dimensional surface and $(n - 2)$-dimensional base characteristics over $[t_2, t_3]$, then it follows from the continuity of $u(t)$ at t_2 that the corresponding integral surface $u_1^{2}(t, x_2^{2})$ has to pass through the initial manifold

$$u_1^{2}(t_2, x_2^{2}) = \begin{pmatrix} u_1^{1}(t_2, g_2(x_2^{2}), x_2^{2}) \\ g_2(x_2^{2}) \end{pmatrix}$$

In this way we can imbed the solution of (4.1.1) over each subinterval $[t_i, t_{i+1}]$ into an i-dimensional integral surface. If the surface $u_1^{n-1}(t, x_n)$ is known over $[t_{n-1}, t_n]$, then the boundary condition

$$f_n(u_1^{n-1}(t_n, x_n), x_n) = 0$$

is a scalar equation in x_n. If a root $x_n = g_n$ can be found, then the n initial values $(u_1^{n-1}(t_n, g_n), g_n)$ are consistent with the boundary values (4.1.2). We shall summarize this discussion as Theorem 4.1.1, which in effect is one recipe for solving multipoint boundary value problems by invariant imbedding.

THEOREM 4.1.1. Assume that for $i = 1, \ldots, n - 1$ the Cauchy problem*

$$\partial u_1^{i}/\partial t + (\partial u_1^{i}/\partial x_2^{i})F_2^{i}(t, u_1^{i}, x_2^{i}) = F_1^{i}(t, u_1^{i}, x_2^{i})$$

$$u_1^{i}(t_i, x_2^{i}) = \begin{pmatrix} u_1^{i-1}(t_i, g_i(x_2^{i}), x_2^{i}) \\ g_i(x_2^{i}) \end{pmatrix}$$

---

* Because of the many superscripts and subscripts we have chosen the notation $\partial u_1^{i}/\partial x_2^{i}$ instead of the notation $u_{1x_2^{i}}^{i}$ to designate the Jacobi matrix $\partial u_k/\partial x_l$, $k = 1, \ldots, i$, $l = i + 1, \ldots, n$.

has a solution $u_1{}^i(t, x_2{}^i)$ over $[t_i, t_{i+1}] \times E_2{}^i$, where g_i is a solution of

$$f_i(u_1^{i-1}(t_i, g_i, x_2{}^i), g_i, x_2{}^i) = 0$$

Then the characteristic $z(t)$ through the point $(u_1^{n-1}(t_n, g_n), g_n)$ is a solution of (4.1.1) and (4.1.2).

Proof. Over each interval $[t_i, t_{i+1}]$, the characteristic $z(t)$ satisfies its characteristic equation (4.1.1). Moreover, by construction $z_1^{n-1}(t)$ remains imbedded in $u^{n-1}(t, x_n)$ over the base characteristic $z_2^{n-1}(t) \equiv z_n(t)$. Hence it satisfies the boundary condition

$$f_{n-1}(z(t_{n-1})) = 0$$

Again, by construction $z_1^{n-2}(t_{n-1}) = u_1^{n-2}(t_{n-1}, z_2^{n-2}(t_{n-1}))$, so that $z_1^{n-2}(t)$ remains imbedded in $u_1^{n-2}(t, x_2^{n-2})$ over $z_2^{n-2}(t)$. In this manner we can proceed from interface to interface and verify that $z(t)$ satisfies (4.1.2) at $t_i, i = 1, \ldots, n$. ∎

The existence theory for boundary value problems in Section 2.3 would indicate that the hypotheses of the theorem can actually be satisfied provided that the interval of integration is sufficiently small and the boundary conditions (4.1.2) are simple. Let us give a quantitative estimate for a particular case.

THEOREM 4.1.2. Let F be continuously differentiable with respect to u and piecewise continuous in t. Suppose also that $|\partial F_i/\partial u_k| \leq C$ for $1 \leq j$, $k \leq n$. Then the multipoint boundary value problem

$$u' = F(t, u), \qquad u_j(t_j) = \alpha_j, \quad j = 1, \ldots, n$$

has a unique solution whenever

$$|t_{j+1} - t_j| < \hat{t}_{j+1} = \frac{1}{(2n-j)C} \ln\left(1 + \frac{2n-j}{j \max\{1, \phi_j\}}\right)$$

where $\phi_1 = 0$ and

$$\phi_j = \frac{(2n-j) \max\{1, \phi_{j-1}\} \exp[(2n-j)C_1 | t_j - t_{j-1}|]}{(2n-j) - j \max\{1, \phi_{j-1}\} \exp[(2n-j)C | t_j - t_{j-1}|]}$$

Proof. Theorems 1.1.3 and 2.3.2 are invoked. It may be verified that for given $j \geq 1$ the constants a, b, c, d in Theorem 1.1.3 can be expressed as $a = jC$, $b = (n-j)C$, $c = jC$, and $d = (n-j)C$ if the matrices corresponding to F_u, F_x, G_u, and G_x are normed by the maximum row sum norm

(N.1.1). It follows from Theorem 1.1.3 that the integral surface $u_1^j(t, x_2^j)$ exists over the interval

$$| t_{j+1} - t_j | < \hat{t}_{j+1} = \frac{1}{(2n - j)C} \ln\left(1 + \frac{2n - j}{j \max\{1, \| \partial u_1^j(t_j, \cdot)/\partial x_2^j \|\}}\right)$$

The bound on $\| \partial u_1^j(t_j, \cdot)/\partial x_2^j \|$ is obtained from Theorem 2.3.2 as

$$\| u_x(t, \cdot) \| \leq \| u_s \| x_s \| \leq \frac{(k + d) \max\{1, i\} e^{(k+d)t}}{(k + d) - c \max\{1, i\}(e^{(k+d)t} - 1)}$$

where $i = \| u_x(0, \cdot) \|$. In the present case this expression translates into

$$\| u_x^j(t, \cdot) \| \leq \frac{(2n - j) \max\{1, i\} \exp((2n - j)C(t - t_{j-1}))}{(2n - j) - j \max\{1, i\}[\exp((2n - j)C(t - t_{j-1})) - 1]}$$

with $i = \| u_x^j(t_j, \cdot) \|$. By construction

$$u_1^j(t_j, x_2^j) = \begin{pmatrix} u^{j-1}(t_j, x_j, x_2^j) \\ \alpha_j \end{pmatrix}$$

Using the maximum row sum norm, we see that $\| u_x^j(t_j, x) \| \leq \| u_x^{j-1}(t_j, x) \|$ which leads to the above recursion equation for ϕ_j. Uniqueness follows from the uniqueness of the characteristic through the point $(u_1^{n-1}(t_n, \alpha_n), \alpha_n)$. ∎

It is apparent from Theorem 4.1.2 that the multipoint boundary value problem has a solution whenever $| t_j - t_{j-1} |$ is sufficiently small, regardless of the magnitude of the $\{\alpha_j\}$. In a sense, then, this theorem can be regarded as an extension of the standard "continuity in the initial value" theorem in the theory of ordinary differential equations, where now each component is perturbed individually. More important for our presentation is the conclusion that the class of multipoint problems amenable to a solution with invariant imbedding is not void (see N.1.2).

Theorem 4.1.1 is attractive for numerical work because as we sweep from t_1 to t_n, the dimension of the base characteristic is reduced at each boundary point t_i which simplifies the calculation (see the comments of Section 2.4). The price we pay is the solution of the Eq. (4.1.3). Suppose for example that $u_1^1(t, x_2^1)$ has been found over $[t_1, t_2]$ and suppose further that the boundary condition at t_2 is given as

$$f_2(u(t_2)) \equiv u_1(t_2) = \alpha$$

According to Theorem 4.1.1, we have to find a function g_2 depending on

$x_2{}^2$ which solves the equation

$$u_1{}^1(t_2, g_2, x_2{}^2) - \alpha = 0$$

For a nonlinear and generally only discretely defined surface $u_1{}^1(t_2, x_2{}^1)$ the solution $x_2 = g_2(x_2{}^2)$ is probably not obtainable exactly and any approximate solution will introduce errors into the integral surface over the succeeding intervals.

There is an alternative and conceptually simpler formulation of the invariant imbedding equations for multipoint problems which is well suited for boundary conditions of the form

$$f_i(u(t_i)) \equiv u_1(t_i) - g_i(x_2{}^1(t_i)) = 0, \qquad i = 1, \ldots, n \qquad (4.1.4)$$

These boundary conditions typically enter in parameter estimation problem where the component u_1 is observed at n different points. We note that the solution of (4.1.1) and (4.1.4) can be imbedded into the family of $n - 1$ initial value problems

$$
\begin{aligned}
u_1{}^{1\prime} &= F_1{}^1(t, u_1{}^1, x_2{}^1), & u_1{}^1(t_i) &= g_i(s_2{}^1), & i = 1, \ldots, n - 1 \\
x_2{}^{1\prime} &= F_2{}^1(t, u_1{}^1, x_2{}^1), & x_2{}^1(t_i) &= s_2{}^1
\end{aligned}
\qquad (4.1.5)
$$

As $s_2{}^1$ ranges over $E_2{}^1$, Eqs. (4.1.5), when regarded as characteristic equations, generate $n - 1$ integral surfaces $u_1{}^1(t, x_2{}^1)_i$, each of which satisfies the same invariant imbedding equation

$$\frac{\partial u^1}{\partial t^1}(t, x_2{}^1) + \frac{\partial u_1{}^1}{\partial x_2{}^1}(t, x_2{}^1)F_2{}^1(t, u_1{}^1, x_1{}^2) = F_1{}^1(t, u_1{}^1, x_2{}^1)$$

and where $u_1(t, x_2{}^1)_i$ passes through the initial manifold

$$u_1{}^1(t_i, x_2{}^1)_i = g_i(x_2{}^1)$$

At the remaining boundary point t_n, we have to satisfy the boundary condition

$$u_1{}^1(t_n, x_2{}^1)_i = g_n(x_2{}^1), \qquad i = 1, \ldots, n - 1 \qquad (4.1.6)$$

This is an $(n - 1)$-dimensional system in the $n - 1$ unknowns $x_2{}^1 = (u_2(t_n), \ldots, u_n(t_n))$. Suppose that a solution $\hat{x}_2{}^1$ of (4.1.6) can be found. Then all the integral surfaces $u_1{}^1(t, x_2{}^1)_i$, $i = 1, \ldots, n - 1$ have the point $(u_1{}^1(t_n, \hat{x}_2{}^1), \hat{x}_2{}^1)$ in common. The characteristic through this point certainly is a solution of the characteristic equations (4.1.1) and, because it must remain imbedded in the integral surfaces $u_1{}^1(t, x_2{}^1)_i$, it satisfies the initial manifold conditions (4.1.4). In summary, we have proved:

THEOREM 4.1.3. Assume that for $i - 1, \ldots, n - 1$ the Cauchy problems

$$\partial u_1^1/\partial t + (\partial u_1^1/\partial x_2^1) F_2^1(t, u_1^1, x_2^1) = F_1^1(t, u_1^1, x_2^1)$$

$$u_1^1(t_i, x_2^1) = g_i(x_2^1)$$

have solutions $u_1^1(t, x_2^1)_i$ over $[t_i, t_n]$. Assume further that the $(n - 1)$-dimensional system

$$u_1^1(t_n, x_2^1)_i = g_n(x_2^1), \qquad i = 1, \ldots, n - 1$$

has a solution $\hat{x}_2^1$. Then the characteristic through $(u_1^1(t_n, \hat{x}_2^1)_1, x_2^1)$ is a solution of the multipoint boundary value problem (4.1.1) and (4.1.4).

Let us briefly contrast this approach to the method described by Theorem 4.1.1. Clearly, each invariant imbedding surface $u_1^1(t, x_2^1)_i$ is defined over an $(n - 1)$-dimensional region so that a numerical solution would be very demanding in time and storage space. In addition, the $(n - 1)$-dimensional system at the final point t_n may be difficult to solve. On the other hand, since we are now looking for a numerical value, this problem would appear to be easier than determining the functions $g_i(x_2^i)$ at each interface t_i. Furthermore, this formulation is much easier to visualize and should be quite useful for multipoint problems with a free interface and for linear differential equations with one nonlinear boundary condition. Moreover, because the same differential equation is to be solved for each i, although subject to different initial conditions, it is to be expected that coding the algorithm for a computer is easy to carry out. Finally, for linear problems, the resulting Riccati equation is frequently autonomous and subject to the same initial value at k different initial times t_i. In this case the Riccati equation needs to be solved only once instead of k times.

It is not difficult to combine the approaches of Theorems 4.1.1 and 4.1.3. The most suitable algorithm will depend on the complexity of the problem and the structure of the prescribed boundary conditions.

Example 4.1.1. The Blasius Equation

A well-known equation of laminar boundary layer theory is the so-called Blasius equation

$$v''' + v''v = 0 \tag{4.1.7}$$

where v' is the dimensionless velocity in the boundary layer of a flat plate in parallel laminar flow. The solution of this equation subject to the two-point boundary values

$$v(0) = v'(0) = 0, \qquad v'(\infty) = 1.0 \qquad (4.1.8)$$

is well established. For a comprehensive discussion of the model, the assumptions, and the behavior of solutions for the more general related Falkner–Skan equation, we refer to the monograph of Evans (1968). (We remark parenthetically that this boundary value problem can be solved through conversion into initial value problems for ordinary differential equations with group theoretic methods [see Na and Hansen (1968)]. An application of invariant imbedding to (4.1.7) and (4.1.8) will, of course, lead to a nonlinear partial differential equation.)

For a numerical solution of the Blasius equation it generally is assumed that the boundary value $v'(\infty) = 1.0$ can be replaced by $v'(R) = 1.0$ for some $R < \infty$. From a physical point of view this appears to be quite acceptable. However, for the sake of argument we shall assume in this example that instead the velocity v' has been observed at three different locations; we shall use the boundary conditions

$$v'(0) = 0, \qquad v'(1) = 0.425, \qquad v'(2) = 0.8 \qquad (4.1.9)$$

which are roughly in accord with the solution of (4.1.7) and (4.1.8) as drawn in Evans (1968, p. 33).

In order to apply invariant imbedding to (4.1.7) we convert the Blasius equation into a first-order system and single out $v'(t) = u(t)$ as surface characteristic. From the formulation

$$
\begin{aligned}
u' &= w, & u(0) &= 0 \\
w' &= -vw, & u(1) &= 0.425 \\
v' &= u, & u(2) &= 0.8
\end{aligned}
$$

we obtain over $[0, 1]$ the invariant imbedding equation

$$\partial u/\partial t + (\partial u/\partial v)u - (\partial u/\partial w)vw = w, \qquad u(0, v, w) = 0 \quad (4.1.10)$$

Equation (4.1.10) will be integrated over the cube $0 \le t \le 1$, $0 \le v \le V$, $0 \le w \le W$ by placing a regular mesh on the cube and approximating (4.1.10) with

$$
(1/\Delta t)[u_{n,i,j} - u_{n-1,i,j}] + (1/\Delta v)[u_{n,i,j} - u_{n,i-1,j}]u_{n,i,j} - v_i w_j/\Delta w
$$
$$
\times [u_{n-1,i,j} - u_{n-1,i,j-1}] = w_j
$$

where $u_{n,i,j} \approx u(n\,\Delta t, i\,\Delta v, j\,\Delta w)$. For the boundary points $v_i = 0$ and $w_j = 0$ an explicit analog to (4.1.10) was used. The above equation is quadratic in $u_{n,i,j}$ and is readily solvable. Incidentally, if $\partial u/\partial w$ is also approximated at the nth level, the quadratic equation has complex roots unless Δt is very small. The same problem does not occur if this term lags behind. It may be noted that the above difference equation corresponds to one of those hybrid methods referred to in Section 2.4. It worked quite well. Figure 4.1.1 shows the computed surface $u(1, v, w)$.

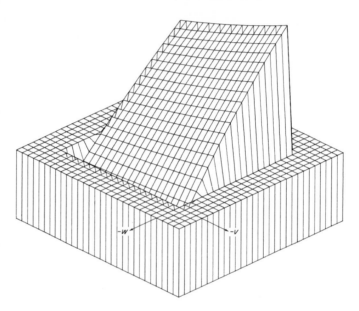

Fig. 4.1.1. Integral surface $u(1, v, w)$ of the boundary layer problem.

It is apparent from this surface that the contour $u(1, v, w) = 0.425$ is readily expressed in the general form $w = f(v)$. Hence it is convenient to consider $\{u(t), w(t)\}$ as the surface characteristic over $[1, 2]$. This characteristic is imbedded into the integral surfaces $u(t, v)$, $w(t, v)$ of

$$\partial u/\partial t + (\partial u/\partial v)u = w, \qquad \partial w/\partial t + (\partial w/\partial v)u = -vw \quad (4.1.11)$$

subject to $u(1, v) = 0.425$, $w(1, v) = f(v)$.

A fully implicit finite difference analog of (4.1.11) allows elimination of one of the dependent variables. The remaining variable satisfies a cubic equation which is readily solved by standard computer routines. Figure 4.1.2 shows the computed curve $u(2, v)$ and the solution point $u(2, v_0) = 0.8$.

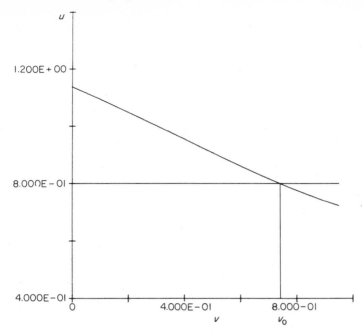

Fig. 4.1.2. Integral surface $u(2, v)$ of the boundary layer problem.

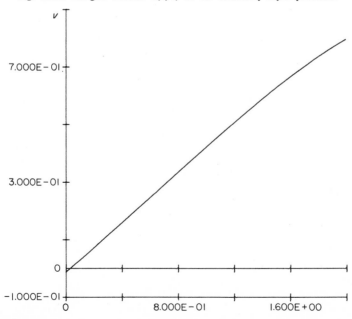

Fig. 4.1.3. Solution $v(t)$ of the Blasius equation obtained by backintegrating the characteristic equations.

Finally, the illustration 4.1.3 contains the solution of the Blasius equation
(4.1.7) over $[0, 2]$ through the initial value $v(2) = v_0$, $u(2) = 0.8$, $w(2)$
$= w(2, v_0)$. This solution was generated with a fourth-order Runge–Kutta
method and matches quite closely the result in Evans for the Blasius equation. It may also be verified that applying the same invariant imbedding
algorithm to the full Falkner–Skan equation will result in quite similar
nonlinear algebraic systems. However, no computations were carried
through for this case. ∎

4.2. LINEAR MULTIPOINT PROBLEMS

The invariant imbedding formulation leading to Theorem 4.1.1 becomes
quite tractable for numerical work if the boundary value problem (4.1.1)
and (4.1.2) is linear. The simplification is, of course, due to the fact that
the surface and base characteristic are related through the affine transformation (1.2.4). Specifically, let us consider the problem

$$u' = A(t)u + F(t)$$
$$u_i(t_i) = f_i u_1^{-1}(t_i) + g_i u_2^i(t_i) + \alpha_i, \qquad i = 1, \ldots, n \tag{4.2.1}$$

where $A(t)$ is an $n \times n$ continuous matrix, where $F(t)$ is a continuous source
function, and where f_i and g_i map E_1^{i-1} and E_2^i into E^1 while α_i is a given
scalar; in other words, the boundary equation (4.1.2) is already solved for
$u_i(t_i)$. The splitting of E^n into $E_1^k + E_2^k$ induces a natural partition of $A(t)$
which we shall write as

$$A(t) = \begin{pmatrix} A_{11}^k(t) & A_{12}^k(t) \\ A_{21}^k(t) & A_{22}^k(t) \end{pmatrix}$$

where $A_{11}^k(t)$ is the $k \times k$ upper left block of $A(t)$. The other submatrices are
defined accordingly. With this notation we can give the invariant imbedding
formulation for (4.2.1) as follows.

Over $[t_1, t_2]$ we need the surface $u_1^1(t, x_2^1)$, which has the representation

$$u_1^1(t, x_2^1) = U_{12}^1(t)x_2^1 + w_1^1(t)$$

where the $1 \times (n - 1)$ matrix U_{12} satisfies the Riccati equation

$$U_{12'}^1 = A_{12}^1(t) + A_{12}^1(t)U_{12}^1 - U_{12}^1 A_{22}^1(t) - U_{12}^1 A_{21}^1(t)U_{12}^1$$
$$U_{12}^1(t_1) = g_1$$

We also verify immediately that w_1^1 is the solution of

$$w_{1'}^1 = [A_{11}^1(t) - U_{12}^1(t)A_{21}^1(t)]w_1^1 + F_1^1(t) - U_{12}^1(t)F_2^1(t)$$

$$w_1^1(t_1) = \alpha_1$$

Let us suppose that $U_{12}^1(t)$ exists on $[t_1, t_2]$. We shall partition $U_{12}^1(t_2)$ into a 1×1 and $1 \times (n-2)$ matrix such that

$$U_{12}^1(t_2)x_2^1 = R_2 x_2 + S_2 x_2^2$$

and express the boundary condition at $t = t_2$ for the component x_2 ($\equiv u_2$) as

$$x_2 = f_2(R_2 x_2 + S_2 x_2^2 + w_1^1(t_2)) + g_2 x_2^2 + \alpha_2$$

This equation has the formal solution

$$x_2 = (1 - f_2 R_2)^{-1}(f_2 S_2 + g_2)x_2^2 + (1 - f_2 R_2)^{-1}(f_2 w_1^1(t_2) + \alpha_2)$$

Setting now

$$D_2 = (1 - f_2 R_2)^{-1}(f_2 S_2 + g_2)$$

$$\beta_2 = (1 - f_2 R_2)^{-1}(f_2 w_1^1(t_2) + \alpha_2)$$

we can write the last equation simply as

$$x_2 = D_2 x_2^2 + \beta_2$$

Hence, the initial condition for $u_1^2(t_1, x_2^2)$ is

$$u_1^2(t_2, x_2^2) = \begin{pmatrix} U_{12}^1(t_2)x_2^1 + w_1^1(t_2) \\ x_2 \end{pmatrix}$$

$$= \begin{pmatrix} (R_2 D_2 + S_2)x_2^2 + R_2 \beta_2 + w_1^1(t_2) \\ D_2 x_2^2 + \beta_2 \end{pmatrix}$$

or

$$u_1^2(t_2, x_2^2) = E_{12}^2 x_2^2 + \gamma_1^2$$

where

$$E_{12}^2 = \begin{pmatrix} R_2 D_2 + S_2 \\ D_2 \end{pmatrix}, \qquad \gamma_1^2 = \begin{pmatrix} R_2 \beta_2 + w_1^1(t_2) \\ \beta_2 \end{pmatrix}$$

Thus, we again have a linear manifold as initial condition so that $u_1^2(t, x_2^2)$ can be found by integrating ordinary differential equations. Quite generally, in order to solve the boundary value problem (4.2.1), we have to solve the following sequence of equations over $[t_k, t_{k+1})$ for $k = 1, \ldots, n - 1$.

$$U_{12'}^k = A_{12}^k(t) + A_{11}^k(t)U_{12}^k - U_{12}^k A_{22}^k(t) - U_{12}^k A_{21}^k(t)U_{12}^k$$
$$U_{12}^k(t_k) = E_{12}^k \tag{4.2.2}$$

$$w_{1'}^k = [A_{11}^k(t) - U_{12}^k(t)A_{21}^k(t)]w_1{}^k + F_1{}^k(t) - U_{12}^k(t)F_{12}^k(t)$$
$$w_1{}^k(t_k) = \gamma_1{}^k \tag{4.2.3}$$

where

$$E_{12}^1 = g_1, \qquad \gamma_1{}^1 = \alpha_1$$

and, for $k \geq 2$,

$$U_{12}^{k-1}(t_k)x_2^{k-1} = R_k x_k + S_k x_2{}^k$$
$$D_k = (1 - f_k R_k)^{-1}(f_k S_k + g_k)$$
$$\beta_k = (1 - f_k R_k)^{-1}(f_k w_1^{k-1}(t_k) + \alpha_k)$$
$$E_{12}^k = \begin{pmatrix} R_k D_k + S_k \\ D_k \end{pmatrix}, \qquad \gamma_1{}^k = \begin{pmatrix} R_k \beta_k + w^{k-1}(t_k) \\ \beta_k \end{pmatrix}$$

and

$$S_n = g_n = 0$$

The desired initial value for $u(t)$ which is consistent with the given boundary values is found from $u(t_n) = \gamma_1{}^n$. The dimensions of the above matrices and vectors are as follows:

$$
\begin{array}{llll}
g_k: & k \times (n - k), & \gamma_1{}^k: & k \times 1, \\
R_k: & (k - 1) \times 1, & S_k: & (k - 1) \times (n - k), \qquad k \geq 2 \\
D_k: & 1 \times (n - k), & E_{12}^k: & k: (n - k) \\
\beta_k: & 1 \times 1 & f_k: & 1 \times (k - 1)
\end{array}
$$

In practice, more than one boundary value may be given at a particular location t_i, as is the case when $t_i = t_{i+j}$ for some j. It now is advantageous to group the components $(x_i, x_{i+j})^T = y_i$ and consider the (two-dimensional) vector y_i at t_i as given. Eqs. (4.2.2) and (4.2.3) remain valid, except that the dimensions of the various matrices must be reinterpreted. For example, the matrix f_k must have dimensions $2 \times k - 1$. The other dimensions are adjusted accordingly. In fact, each component u_i of the vector u may itself be an m_i-dimensional vector.

Equations (4.2.2) and (4.2.3) may appear complex but the difficulty is primarily notational. In addition, in most practical problems the boundary conditions are simpler than those of (4.2.1) so that the matrices E_{12}^k and vectors $\gamma_1{}^k$ are readily obtainable.

It is apparent from this discussion that generally we are faced with Riccati equations of different row and column size. Little information beyond Theorems 1.3.1 and 1.3.2 is presently available for such nonsquare systems and their solution can only be assumed to exist. For diffusion-type problems, this assumption is borne out by numerical calculations. If, however, the Riccati solution has an unbounded solution, the straightforward switching of the definition of the surface and base characteristic used in Section 3.5 is impossible because inverses are not defined. There is, of course, no difficulty in deriving the Riccati equation when surface and base characteristics are reversed. The problem arises in defining the proper initial values at the switching interfaces.

The invariant imbedding technique is designed to produce the initial value $u(t_n)$ which is consistent with the given boundary values. There still remains the problem of integrating (4.2.1) subject to the computed initial value. We have seen examples in previous sections where numerical instabilities precluded a direct solution of the initial value problem. Instead, a backward sweep was employed. The same approach is also applicable to multipoint problems. Suppose, for example, that the functions $U_{12}^k(t)$, $w_1{}^k(t)$ are available over $[t_k, t_{k+1}]$ for $k = 1, \ldots, n - 1$. Given $x_2^{n-1}(t_n)$ [i.e., the nth component of the computed initial value $u(t_n)$], we can find $u(t)$ over $[t_{n-1}, t_n]$ by integrating

$$x_2^{n-1} = [A_{21}^{n-1}(t)U_{12}^{n-1}(t) + A_{22}^{n-1}(t)]x_2^{n-1} + A_{21}^{n-1}(t)w_1^{n-1}(t) + F_2^{n-1}(t)$$

and setting

$$u_1^{n-1}(t) = U_{12}^{n-1}(t)x_2^{n-1}(t) + w_1^{n-1}(t)$$

With $x_2^{n-2}(t_{n-1})$ given by these expressions, we then integrate the differential equation for $x_2^{n-2}(t)$ over $[t_{n-2}, t_{n-1}]$. In this way, the solution is pieced together interval by interval. It seems reasonable to assume that stability properties observed for two-point problems have their analogs for multipoint problems. However, it is also apparent that this reverse sweep will generally lead to considerable bookkeeping difficulties when the algorithm is implemented on the computer.

As before, it should be emphasized that the same reduction to initial value problems for ordinary differential equations remains valid when one of the boundary conditions is nonlinear or when two boundary conditions are specified at a free interface. In this case the imbedding equations are integrated from the left and right toward the nonlinearity to obtain the affine transformation at the troublesome boundary point. Once the transformation is known, a nonlinear algebraic system will result. This approach would be the analog of Theorem 4.1.3 for linear problems. An illustration of this technique is given in Example 4.2.2.

Example 4.2.1. The Potential Equation on an Annulus

In order to illustrate the applicability of Eqs. (4.2.2), and (4.2.3) let us consider the following boundary value problem for Laplace's equation

$$
\begin{aligned}
\Delta u &= 0, & (r, \theta) &\in D \\
\partial u / \partial r &= 1, & (r, \theta) &\in \partial D_1 \\
u &= f(\theta), & (r, \theta) &\in \partial D_2
\end{aligned}
\tag{4.2.4}
$$

where D, ∂D_i are shown as in Fig. 4.2.1. Generally, $f(\theta)$ is an arbitrary function of period 2π, although for the specific example discussed here it is assumed to be symmetric about the points $\theta = 0$ and $\theta = \pi/2$.

Conventional finite difference schemes for this problem are awkward because of the curvilinear and straight boundaries. We shall attack this problem with the method of lines and discretize the angle θ. Because of the assumed symmetry of f, it suffices to consider only the segment $0 \le \theta \le \pi/2$

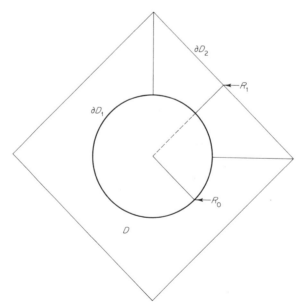

Fig. 4.2.1. Annular domain for the diffusion equation.

this simplification will reduce the size of the resulting system of ordinary differential equations. Denoting by u_i the solution of (4.2.4) along the ray $\theta_i = (i - 1) \Delta\theta$ for some increment $\Delta\theta = \pi/2(N - 1)$, $N > 1$, we can replace Laplace's equation with the system

$$u_i'' + \frac{1}{r} u_i' + \frac{1}{r^2} \frac{u_{i+1} + u_{i-1} - 2u_i}{\Delta\theta^2} = 0, \qquad i = 1,\ldots, N \quad (4.2.5)$$

where, because of symmetry,

$$u_{-1}(r) = u(r), \qquad u_{N+1}(r) = u_{N-1}(r)$$

The boundary conditions are

$$u_i'(R_0) = 1; \quad u_i(r_i) = f(\theta_i)$$

where, by Fig. 4.2.1,

$$r_i = R_1/\cos(\pi/4 - \theta_i) \tag{4.2.6}$$

For ease of exposition, we shall choose $N = 3$ so that we are concerned only with the rays

$$\theta_1 = 0, \quad \theta_2 = \pi/4, \quad \theta_3 = \pi/2$$

We shall not specify f beyond requiring (for ease of computation) that

$$f(0) = 1, \quad f(\pi/4) = 0, \quad f(\pi/2) = \tfrac{1}{2}$$

As usual, the second-order equation (4.2.5) is converted into a first-order system. To be consistent with the notation introduced above, we shall use

$$v_1 = u_1', \quad v_2 = u_2', \quad v_3 = u_3', \quad v_4 = u_1, \quad v_5 = u_2, \quad v_6 = u_3$$

The differential equations can be rewritten for $v = (v_1,\ldots, v_6)^T$ as

$$v' = A(r)v \tag{4.2.7}$$

where

$$-A(r) = \begin{pmatrix} 1/r & 0 & 0 & -2/r^2 \Delta\theta^2 & 1/r^2 \Delta\theta^2 & 0 \\ 0 & 1/r & 0 & 1/r^2 \Delta\theta^2 & -2/r^2 \Delta\theta^2 & 1/r^2 \Delta\theta^2 \\ 0 & 0 & 1/r & 0 & 1/r^2 \Delta\theta^2 & -2/r^2 \Delta\theta^2 \\ 1 & 0 & 0 & 0 & 0 & 0 \\ 0 & 1 & 0 & 0 & 0 & 0 \\ 0 & 0 & 1 & 0 & 0 & 0 \end{pmatrix}$$

subject to the boundary conditions

$$v_1 = v_2 = v_3 = 1, \qquad \text{at} \qquad r = R_0$$

$$v_4(R_1 \sqrt{2}) = 1, \quad v_5(R_1) = 0, \quad v_6(R_2 \sqrt{2}) = \tfrac{1}{2}$$

In this exposition we shall integrate successively over the intervals $[\xi_1, \xi_4]$, $[\xi_4, \xi_5]$, $[\xi_5, \xi_6]$, where $\xi_1 = R_0$, $\xi_4 = r_1$, $\xi_5 = r_2$, $\xi_6 = r_3$, with r_i given by (4.2.6). Because the initial conditions are given for v_1, v_2, v_3, we start with the integral surface

$$v_1{}^3(r, x_2{}^3) = U_{12}^3(r)x_2{}^3 + w_1{}^3(r)$$

We recall that $x_2{}^3 = (v_4, v_5, v_6)$ designates those components of v which are taken as base characteristics. Since $v_1{}^3(\xi_1, x_2{}^3) = (1, 1, 1)^{\mathrm{T}}$ for arbitrary $x_2{}^3$, it follows that $U_{12}^3(\xi_1) = E_{12}^3 = 0$ and $w_1{}^3(\xi_1) = \gamma_1{}^3 = (1, 1, 1)^{\mathrm{T}}$. This will start Eqs. (4.2.2) and (4.2.3) at $\xi_1 = R_0$.

We integrate (4.2.2) and (4.2.3) for $k = 3$ subject to the given initial values. From the boundary condition

$$v_4(\xi_4) = x^4 = 1$$

we find that $f_4 = 0$, $g_4 = 0$, and $\alpha_4 = 1$. Hence

$$R_4 = \begin{pmatrix} (U_{12}^3(\xi_4))_{14} \\ (U_{12}^3(\xi_5))_{24} \\ (U_{13}^3(\xi_6))_{34} \end{pmatrix}$$

where $(U_{12}^3(\xi_4))_{ij}$ is the element in the ith row and jth column of the 6×6 matrix U. (Of course, the components U_{11}^3, U_{21}^3, U_{22}^3 are not even computed. The above notation is chosen solely to maintain a regularity for subscripting which is helpful in computer applications.) Similarly,

$$S_4 = \begin{pmatrix} (U_{12}^3(\xi_4))_{15} & (U_{12}^3(\xi_4))_{16} \\ (U_{12}^3(\xi_4))_{25} & (U_{12}^3(\xi_4))_{26} \\ (U_{12}^3(\xi_4))_{35} & (U_{12}^3(\xi_4))_{36} \end{pmatrix}$$

$$D_4 = (0 \quad 0), \qquad \beta_4 = 1$$

These quantities define E_{12}^4 and $\gamma_1{}^4$ and allow us to integrate the (4×2)-dimensional Riccati equation for U_{12}^4 and the linear equation (2.4.3) (backward) over $[R_1, R_1 \sqrt{2}]$. At ξ_5 we find that

$$
R_5 = \begin{pmatrix} (U_{12}^4(\xi_5))_{15} \\ (U_{12}^4(\xi_5))_{25} \\ (U_{12}^4(\xi_5))_{35} \\ (U_{12}^4(\xi_5))_{45} \end{pmatrix}, \qquad S_5 = \begin{pmatrix} (U_{12}^4(\xi_5))_{16} \\ (U_{12}^4(\xi_5))_{26} \\ (U_{12}^4(\xi_5))_{36} \\ (U_{12}^4(\xi_5))_{46} \end{pmatrix}
$$

$$
D_5 = 0, \qquad \beta_5 = 0
$$

From these quantities we obtain E_{12}^5 and γ_1^5 and integrate $U_{12}^5(r)$ and $w_1^5(r)$ over $[\xi_5, \xi_6]$. The final solution $v(R_1 \sqrt{2}')$ is given by

$$
v(R_1 \sqrt{2}) = \gamma_1^6 = \left(\begin{pmatrix} (U_{12}^5(\xi_6))_{16} \\ (U_{12}^5(\xi_6))_{26} \\ (U_{12}^5(\xi_6))_{36} \\ (U_{12}^5(\xi_6))_{46} \\ (U_{12}^5(\xi_6))_{56} \end{pmatrix} \times \tfrac{1}{2} + v_1^5(\xi_6) \right) \bigg/ \tfrac{1}{2}
$$

Note that $v(R_1 \sqrt{2})$ is actually the continuation of the solution of (4.2.4) to the circle circumscribing D.

Problem (4.2.4) was solved numerically along the three rays indicated. However, to cut down on the number of intervals, the components of u were grouped as follows

$$
v = (u_1', u_2', u_3', u_5, (u_4, u_6)^{\mathrm{T}})^{\mathrm{T}}
$$

where (u_4, u_6) was treated as one component with boundary value $(1, \tfrac{1}{2})$. This reduced the four point formulation (2.4.7) to a three-point problem. Of course, this simplification required rearrangement of the matrix $A(r)$. As usual, a fourth-order Runge–Kutta method was used for the numerical integrations. Table 4.2.1 lists the parameters employed and the values obtained.

Example 4.2.2 is one of those cases where heuristic arguments may be invoked to insure the existence of solutions for the Riccati equation. If the outer boundary is circular as well, then the invariant imbedding formulation will lead to a two-point boundary value problem. Moreover, it is straightforward to show that the equations obtained from discretizing the Laplacian can be arranged in such an order that the matrix Riccati equation is Hermitian with a bounded positive definite solution. If, as is usually assumed, invariant imbedding is to reflect physical reality, then our multipoint problem can be viewed as a perturbation of the two-point problem,

Table 4.2.1

Numerical Solution of the Multipoint Problem $(4.2.7)^a$

	$\xi = 1$	$\xi = 2$	$\xi = 2\sqrt{2}$
Computed solution			
u_1	-0.491^b	0.379^b	0.176
u_2	-0.600^b	-0.003^b	0.5
u_3	-0.595^b	0.098^b	1
u_1'	0.984^b	0.753^b	0.761
u_2'	0.985^b	0.353^b	0.096
u_3'	0.995^b	0.536^b	0.456
Difference between the computed and given boundary values			
u_1	—	—	—
u_2	—	0.003	0
u_3	—	—	0
u_1'	0.016	—	—
u_2'	0.015	—	—
u_3'	0.005	—	—

^{a} Parameters: $R_0 = 1$; $R_1 = 2$; $\Delta\xi = 10^{-2}$.
^{b} Denotes values obtained from a backintegration of (2.4.5) subject to the computed initial value at $\xi = 2\sqrt{2}$ (N.2.1). ■

where each boundary point is varied individually. Laplace's equation will have a solution over any annulus like domain provided the boundaries and boundary values are sufficiently smooth. Invariant imbedding should account for this absence of critical lengths which in turn would require bounded solutions for the Riccati equation.

Example 4.2.2. Deflection of an Elastic Rail

The following example from the field of engineering statics will illustrate use of the alternative Theorem 4.1.3 for the solution of a multipoint problem.

The transverse displacement of an elastically imbedded rail subject to a distributed transverse load is described by the (linear) fourth-order equation

$$(EJ(\xi)u'')'' + K(\xi)u = g(\xi)$$

where $EJ(\xi)$ is the flexural rigidity, $K(\xi)$ the elastic resistance of the supporting material, and $g(\xi)$ the load density. For a freely supported rail the boundary conditions are $u''(-L) = u'''(-L) = u''(L) = u'''(L) = 0$ and correspond to vanishing moments and shear forces at the rail ends. This problem is discussed and solved in Collatz (1960, p. 152). We shall use the same differential equation. However, it will be assumed that the rail is hinged in a complicated (meaning not analyzable) manner at its endpoints and that the moments and shear forces have to be determined at the rail ends from the measured displacements at four different points along the rail.

When EJ, g, and K are given by

$$EJ(\xi) = EJ_0(2 - (\xi/L)^2), \qquad g(\xi) = g_0(2 - (\xi/L)^2)$$

the equation for the elastic rail can be converted into the nondimensional form

$$[(2 - t^2)v'']'' + kv = 2 - t^2$$

where $t = \xi/L$, $v = (EJ_0/g_0L^4)u$, $k = (L^4/EJ_0)K$. As in Collatz (1960), we shall assume that $k = 40$.

By defining z through the relation

$$z = (2 - t^2)v''$$

the above equation can further be reduced to

$$z'' + 40v - 2 + t^2 = 0, \qquad v'' - z/2 - t^2 = 0 \qquad (4.2.8)$$

We shall assume that the following (dimensionless) displacements were observed.

$$t_1 = 0.2, \qquad v(t_1) = \alpha_1 = 0.0448156$$
$$t_2 = 0.4, \qquad v(t_2) = \alpha_2 = 0.0433224$$
$$t_3 = 0.6, \qquad v(t_3) = \alpha_3 = 0.0410152 \qquad (4.2.9)$$
$$t_4 = 0.8, \qquad v(t_4) = \alpha_4 = 0.0381534$$

If we let

$$y = \begin{pmatrix} v \\ v' \\ z \\ z' \end{pmatrix}$$

the system (2.4.8) becomes

$$y' = A(t)y + F(t)$$

where

$$A(t) = \begin{pmatrix} 0 & 1 & 0 & 0 \\ 0 & 0 & 1/(2 - t^2) & 0 \\ 0 & 0 & 0 & 1 \\ -40 & 0 & 0 & 0 \end{pmatrix}, \qquad F(t) = \begin{pmatrix} 0 \\ 0 \\ 0 \\ 2 - t^2 \end{pmatrix}$$

and the boundary conditions can be written as

$$y_1(t_i) = \alpha_i$$

where t_i and α_i are given in (2.4.9). According to Section 4.1, this system is to be solved by finding the integral surfaces $y_1^1 (t, x_2^1)_i$ through the initial manifolds $y_1^1(t_i, x_2^1)_i = \alpha_i$, $i = 1, 2, 3$, and by solving for x_2^1 at $t = t_4$ from the three equations

$$y_1^1(t_4, x_2^1)_i = \alpha_4, \qquad i = 1, 2, 3 \tag{4.2.10}$$

Because Eqs. (2.4.8) are linear, we can write

$$y_1^1(t_4, x_2^1)_i = U_{12}^1(t_4)_i x_2^1 + w_1^1(t_4)_i \tag{4.2.11}$$

where $U_{12}^1(t)_i$ and $w_1^1(t)_i$ satisfy the initial value problems

$$U_{12'}^1(t)_i = A_{12}^1(t) + A_{11}^1(t)U_{12}^1(t)_i - U_{12}^1(t)_i A_{22}^1(t) - U_{12}^1(t)_i A_{21}^1(t)U_{12}^1(t)_i$$
$$U_{12}^1(t_i)_i = 0$$
$$w_1^1(t)_i' = [A_{11}^1(t) - U_n^1(t)_i A_{21}^1(t)]w_1^1(t)_i - U_{12}^1(t)_i F_2^1(t) + F_1^1(t)$$
$$w_1^1(t_i)_i = \alpha_i$$

$$\tag{4.2.12}$$

We note that

$$A_1^1(t) = 0, \qquad A_{12}^1(t) = (1 \quad 0 \quad 0), \qquad A_{21}^1 = \begin{pmatrix} 0 \\ 0 \\ -40 \end{pmatrix},$$

$$A_{22}^1 = \begin{pmatrix} 0 & 1/(2 - t^2) & 0 \\ 0 & 0 & 1 \\ 0 & 0 & 0 \end{pmatrix}$$

$$F_1^1(t) = 0 \quad \text{and} \quad F_1^1(t) = \begin{pmatrix} 0 \\ 0 \\ 2 - t^2 \end{pmatrix}$$

Equations (2.4.12) were integrated as a system with the Runge–Kutta method with step size $\Delta t = 10^{-3}$. The resulting affine transformation (2.4.11)

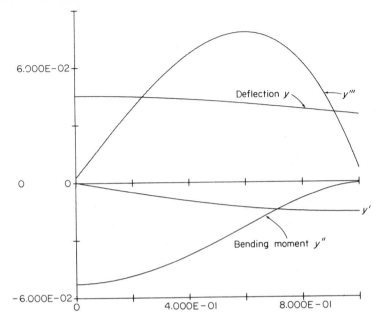

Fig. 4.2.2. Deflection of an elastic rail.

was substituted into the boundary condition (2.4.10) and solved for $x_2^1(t_4)$. Gaussian elimination was used. The vector $(\alpha_4, x_2^1(t_4))$ served as initial value for integrating the system (4.2.8) forward and backward over $[0, 1]$ with a fourth-order Runge–Kutta method. At $t = 1$, the following values were obtained.

$$y_1(1) = 0.035058 \qquad (0.0350551)$$
$$y_2(1) = -0.015739$$
$$y_3(1) = 0.000503 \qquad (0.0)$$
$$y_4(1) = 0.007199 \qquad (0.0)$$

where the values in parentheses are the numerical results given by Collatz. Figure 4.2.2 shows the four components of $y(t)$ over $[0, 1]$. There is good agreement with the values of Collatz.

It was observed that this problem had a solution for sufficiently small k only. A look at the Riccati equation (4.2.12) shows that this is not surprising. Writing out the equations for U_{12}^1 we obtain the three-dimensional system

$$U_1' = 1 + kU_1U_3, \qquad U_1(t_i) = 0$$
$$U_2' = \{1/(2 - t^2)\}U_1 + kU_2U_3, \qquad U_2(t_i) = 0$$
$$U_3' = -U_2 + kU_3^2, \qquad U_3(t_i) = 0$$

We see that $U_1{}'(0) > 0$, $U_2{}''(0) < 0$ and $U_3{}'''(0) > 0$; hence U_2 is concave and U_3 is convex over some neighborhood $(0, \varepsilon)$. Moreover, U_1 will remain positive, forcing U_2 to stay negative and U_3 to remain positive. Hence U_3 will grow quadratically and eventually blow up. It is well known that the related simpler fourth-order equation $v^{(iv)} + \lambda v = 0$ for $\lambda > 0$ leads to an eigenvalue problem (Kamke, 1948, p. 525). Hence it is to be expected that critical length considerations apply to (4.2.8). ∎

We hope that it is apparent to the reader at this point how the characteristic theory approach and the imbedding into initial value problems can be adapted to multipoint boundary value problems with additional fixed and a free interface, how linear equations with one nonlinear fixed or free interface or boundary are to be treated, or how a dynamic programming equation can be derived for control problems where the state or cost equation is subject to multipoint boundary conditions.

To what extent invariant imbedding is applicable to the most general multipoint problem of the form

$$y' = H(t, y), \qquad f(y(t_1), \ldots, y(t_m)) = 0$$

is not yet clear.

NOTES

N.1.1. Throughout Chapter 1 the use of the Euclidean norm was implied in our discussion. However, it is readily verified that almost all results of that chapter remain valid for arbitrary norms on the spaces containing the surface and base characteristics. The only exception is Lemma 1.1.1 where the differentiability of the norm was required.

N.1.2. A somewhat different inequality for $|t_{j+1} - t_j|$ was presented in Meyer (1970a). This bound was based on estimating the invertibility of x_s from the integral equation

$$x_s = I + \int_0^t (G_u u_s + G_x x_s)\, dr$$

by assuring that the norm of the integral was less than 1. This leads to a cruder bound on the permissible $\hat{t}$ than that given in Theorem 1.1.3.

N.2.1. This example is meant to illustrate the invariant imbedding approach. If the actual solution is of interest the reverse sweep should be used since a pure initial value problem for Eq. (4.2.5) has exponentially growing solutions which may cause stability problems.

Chapter

5

Invariant Imbedding for Abstract Equations

5.1. DIFFERENTIAL EQUATIONS IN A BANACH SPACE

It may be noted from Chapter 1 that the equivalence between characteristic equations and the corresponding invariant imbedding equation is based on nothing other than the chain rule for differentiation. Since it is well known that the chain rule holds for differentiable Banach space valued functions, it may be expected that the theory of invariant imbedding has a natural extension to boundary value problems for abstract differential equations (see N.1.1). Of course, such a theory will primarily be of analytical rather than computational value (although several algorithms based on approximating the abstract equations have been used and are briefly described below). However, it does provide a unifying view of the many known invariant imbedding equations for partial and integro-differential equations. In particular, the theory provides a tool for deriving invariant imbedding equations for radiative transfer processes directly from the classical Boltzmann transport equations under clearly discernable assumptions and thus allows us to recover the results on particle transport processes which form the core of the traditional invariant imbedding literature [see, e.g., Bailey (1964)].

Before we can develop the characteristic theory and apply it to the conversion of boundary value problems into initial value problems, we need a number of definitions and preliminary results arising in the study of differential equations in a Banach space. All the concepts used in this section are quite elementary and little different from those of the previous chapters. In a sense, then, this section provides a review of much of the material presented so far.

Throughout this chapter we will deal with Frechet differentiable, Banach space valued functions and abstract differential equations. Let us define these terms and collect some well-known results for coordinate-free differential equations. A more detailed discussion of abstract functions may be found in Vainberg (1964). A coordinate free presentation of the classical theory of ordinary differential equations is given in Dieudonné (1969).

Subsequently, X_i, $i = 1, 2, \ldots$, will denote arbitrary Banach spaces with elements x_i and norm $\| \ \|_i$. Frequently, the use of X and Y will be convenient instead of X_i; D_i is an open subset of X_i while the space of bounded linear operators from X_i to X_j with the uniform operator topology is denoted by $L(X_i, X_j)$. In other words, if $A \in L(X_i, X_j)$ then

$$\| A \|_{ij} = \sup \{ \| Ax_i \|_j \colon \| x_i \|_i = 1 \}.$$

When the meaning is clear, we shall omit the subscripts i, j in order to simplify the notation. We also shall reserve the parameters t, ξ, z for the real valued independent variable which is restricted to a specified interval $R \subset (-\infty, \infty)$.

DEFINITION 5.1.1. An operator acting from an n-dimensional Euclidean space into an arbitrary Banach space is called an *abstract function*.

In this discussion we shall always restrict ourselves to the case $n = 1$ so that an abstract function typically will be written as $x(t)$. The derivative of such an abstract function is defined in the usual manner as

$$x'(t) = \lim_{\Delta t \to 0} \{ x(t + \Delta t) - x(t) \} / \Delta t$$

A somewhat more general concept of differentiation is needed for functions defined on a Banach space.

DEFINITION 5.1.2. A function $F \colon D_1 \subset X_1 \to X_2$ is *Frechet differentiable* at a point $x_0 \in D_1$ if there exists an operator (denoted by) $F_x(x_0) \in L(X_1, X_2)$ such that

$$\lim_{\|h\|_1 \to 0} \| F(x_0 + h) - F(x_0) - F_x(x_0)h \|_2 / \| h \|_1 = 0$$

for any $h \in X_1$.

In this terminology, the derivative of an abstract function given above is a bounded linear operator from the real line to a Banach space and is defined by $x'(t_0): \Delta t \to x'(t_0) \Delta t$ for all Δt. If X_1 and X_2 are Euclidean spaces then F_x has a standard matrix representation and corresponds to the derivative used in the preceding chapters. We shall say that a Banach space valued function F is continuously differentiable if $F_x(x)$ is continuous in x.

DEFINITION 5.1.3. An *abstract ordinary differential* equation is a differential equation whose solution is an abstract function.

We generally will write abstract equations in the usual differential equation form

$$x' = F(t, x) \tag{5.1.1}$$

where $F: R \times D \to X$. A concrete example may be given as

$$\partial x(t, r)/\partial t = \int_a^b f(t, r, s) x^2(t, s) \, ds \tag{5.1.2}$$

where $t \in [0, 1]$, $r \in [a, b]$, where, for fixed t, $x(t, r)$ is an element of the space $C[a, b]$ of continuous functions on $[a, b]$ with norm $\| x(t, \cdot) \| = \sup\{| x(t, r) |: a \le r \le b\}$ and where f is a continuous function on $[0, 1] \times [a, b] \times [a, b]$. To indicate how this equation represents a generalization of the material of the preceding chapters, we note that if we were to approximate x with the n-dimensional vector

$$x = (x(r_1), x(r_2), \ldots, x(r_n))$$

where $a = r_1 \le r_2 \le \cdots \le r_n = b$ is a partition of $[a, b]$, and to evaluate the integral with the trapezoidal rule, then (5.1.2) can be approximated by the n-dimensional system

$$x'(r_i) = \tfrac{1}{2} f(t, r_i, a) x^2(a) + \sum_{j=2}^{n-1} f(t, r_i, r_j) x^2(r_j)$$
$$+ \tfrac{1}{2} f(t, r_i, b) x^2(b), \qquad i = 1, \ldots, n$$

Often it is useful to consider differential equations in a Banach space setting when the right hand side of (5.1.1) depends on a parameter (like r in (5.1.2)).

Modern existence and uniqueness theory for differential equations is based on fixed point theory in coordinate-free spaces (Dieudonné, 1969) and applies to classical and abstract equations alike. Consequently, the well known theorems of standard ordinary differential equations theory apply to the abstract equation (5.1.1), (but not the Cauchy-Peano existence theorem for differential equations with only continuous right-hand sides (Dieudonné, 1969, p. 290)). In particular, the following results are pertinent to our work.

THEOREM 5.1.1. Let $F: R \times D \subset R \times X \to X$ be continuous in t and Lipschitz continuous in x. Then the initial value problem

$$x' = F(t, x), \qquad x(t_0) = x_0, \quad (t_0, x_0) \in R \times D$$

has a unique solution $x(t, t_0, x_0)$ near t_0 which is uniformly continuous in (t, t_0, x_0).

THEOREM 5.1.2. Let F be p-times continuously differentiable ($p \geq 1$) in t and x, then the solution $x(t, t_0, x_0)$ is p-times continuously differentiable with respect to (t, t_0, x_0).

We note that it was precisely Theorem 5.1.2 which in Chapter 1 allowed construction of the integral surface for the invariant imbedding equation by integrating the characteristic equations subject to a free parameter. Thus, there is no difficulty in generating an integral surface through a given initial manifold from the characteristics. Moreover, it is readily verified that the quantitative estimates of Theorem 1.1.3 depend on norms only and are independent of the underlying spaces. Without further ado then we can state the analog of Theorem 2.3.1 for abstract two-point boundary value problems defined on the interval $R = (0, T)$.

THEOREM 5.1.3. Suppose that $F: R \times X_1 \times X_2 \to X_1$ and $G: R \times X_1 \times X_2 \to X_2$ are continuous in t and continuously differentiable in x_1 and x_2. Assume further that there exist constants a, b, c, d, and i such that $\| F_{x_1} \| \leq a$, $\| F_{x_2} \| \leq b$, $\| G_{x_1} \| \leq c$, $\| G_{x_2} \| \leq d$, $\| f_{x_2} \| \leq i$ uniformly in x_1, x_2, and t. Then the abstract boundary value problem

$$x_1' = F(t, x_1, x_2), \qquad x_1(0) = f(x_2)$$
$$x_2' = G(t, x_1, x_2), \qquad x_2(T) = \hat{b}$$

has a unique solution provided

$$T < \hat{\imath} = 1/(k + d) \ln(1 + (k + d)/c \max\{i, 1\})$$

where $k = \max\{a + b, c + d\}$. Moreover, this solution is the characteristic

$\{x_1(t), x_2(t)\}$ through the point $(x_1(T, \hat{b}), \hat{b}) \in X_1 \times X_2$ where $x_1(t, x_2)$ is the integral surface of*

$$\partial x_1/\partial t + (x_1)_{x_2} G(t, x_1, x_2) = F(t, x_1, x_2)$$

$$x_1(0, x_2) = f(x_1)$$

Similarly, all of the other material of the preceding chapters can be rephrased for abstract equations with differentiable right-hand sides. As a consequence, we obtain a number of existence and uniqueness theorems for abstract two-point boundary value problems, interface problems, multi-point problems and free boundary value problems.

In invariant imbedding the generation of $u(t, x)$, usually without use of the characteristics, is the first step of the solution algorithm for boundary value problems. The difficulties arising in establishing the existence of the integral surface $u(t, x)$ without use of the characteristic equations were already commented upon in Section 1.4. Undoubtedly, they are magnified in an infinite dimensional setting. So from a practical point of view, the theory, while applicable to general nonlinear boundary value problems, will yield only useful existence and uniqueness theorems if the problem is close to linear and the interval of integration $(0, T)$ is sufficiently small so that Theorem 5.1.3 holds.

On occasion, however, it may be possible to guess a functional representation for $x_1(t, x_2)$ in the manner indicated in Section 1.4, which reduces the invariant imbedding equation to a more manageable form. As an example, which at the same time puts into a more concrete setting the abstract theory, suppose that we are confronted with the boundary value problem

$$(\partial x_1/\partial t)(t, r) = \int_a^b A(t, r, s)x_1(t, s) \, ds + \int_a^b B(t, r, s)x_2{}^2(t, s) \, ds$$

$$(\partial x_2/\partial t)(t, r) = D(t, r)x_2(t, r) \tag{5.1.3}$$

$$x_1(0, r) = \int_a^b f(r, s)x_2{}^2(0, s) \, ds, \quad g(x_2(T, r), x_2(T, r)) = 0$$

where, for simplicity, A, B, and D are taken to be continuous in all arguments. We shall look for a solution $\{x_1(t, \cdot), x_2(t, \cdot)\}$ in the space $L_2[a, b]$ of square integrable functions defined on $[a, b]$. Equations (5.1.3) can be

---

* A word about notation: $x_1(t, x_2)$ for fixed x_2 is an abstract function in t and $\partial x/\partial t$ denotes its derivative, while for fixed t the expression $(x_1)_{x_2}$ is the Frechet derivative of x_1 with respect to x_2.

interpreted as an abstract boundary value problem by writing

$$x_1' = \tilde{A}(t)x_1 + \tilde{B}(t, x_2), \qquad\qquad x_1(0) = \tilde{f}(x_2(0))$$
$$x_2' = \tilde{D}(t)x_2, \qquad\qquad g(x_1(T), x_2(T)) = 0$$

where $\tilde{B}$ and $\tilde{f}$ are nonlinear operators defined on $L_2[a, b]$ by

$$\tilde{B}(t, x_2) = \int_a^b B(t, \cdot, s)x_2^2(s)\, ds, \qquad \tilde{f}(x_2) = \int_a^b f(\cdot, s)x_2^2(s)\, ds$$

$\tilde{A}$ and $\tilde{D}$ are the linear operators defined on $L_2[a, b]$ by

$$\tilde{A}(t)x_1 = \int_a^b A(t, \cdot, s)x_1(s)\, ds, \qquad \tilde{D}(t)x_2 = D(t, \cdot)x_2(\cdot)$$

The corresponding invariant imbedding equation is

$$\partial x_1/\partial t + (x_1)_{x_2}\tilde{D}(t)x_2 = \tilde{A}(t)x_1 + \tilde{B}(t, x_2), \qquad x_1(0, x_2) = \tilde{f}(x_2)$$

where for fixed (t, x) the operator $(x_1)_{x_2}(t, x_2)$ belongs to the space $L(L_2[a, b]$, $L_2[a, b])$. Concretely, this imbedding equation can be written as

$$\partial x_1/\partial t + (x_1)_{x_2}D(t, r)x_2(r) = \int_a^b A(t, r, s)x_1(t, s)\, ds$$
$$+ \int_a^b B(t, r, s)x_2^2(s)\, ds$$
$$x_1(0, x_2) = \int_a^b f(r, s)x_2^2(s)\, ds$$

Note that t is a scalar and x_2 an L_2-valued independent variable.
In analogy to the finite dimensional case we guess a representation for $x_1(t, x_2)$ of the following form

$$x_1(t, x_2) = \int_a^b U(t, r, s)x_2^2(s)\, ds \tag{5.1.4}$$

where U is an as yet unknown kernel.

It is readily verified from the definition that the Frechet derivative of x_1 with respect to x_2 is given as

$$(x_1)_{(x_2)}h(r) = 2\int_a^b U(t, r, s)x_2(s)h(s)\, ds$$

Hence substitution of (5.1.4) into the invariant imbedding equation leads to

$$\int_a^b \frac{\partial U}{\partial t} (t, r, s) x_2{}^2(s) \, ds + 2 \int_a^b (U(t, r, s) x_2(s) D(t, s) x_2(s) \, ds$$

$$= \int_a^b A(t, r, s) \left[\int_a^b U(t, s, z) x_2{}^2(z) \, dz \right] ds + \int_a^b B(t, r, s) x_2{}^2(s) \, ds$$

Relabeling the dummy variables of integration, we can bring this equation into the form

$$\int_a^b \left[\frac{\partial U}{\partial t} (t, r, s) + 2U(t, r, s) D(t, s) \right.$$

$$\left. - \int_a^b A(t, r, z) U(t, z, s) \, dz - B(t, r, s) \right] x_2{}^2(s) \, ds = 0$$

This expression has to hold for arbitrary $x_2 \in L_2[a, b]$ hence the kernel of (5.1.4) must satisfy

$$\frac{\partial U}{\partial t} (t, r, s) + 2U(t, r, s) D(t, s) - \int_a^b A(t, r, z) U(t, z, s) \, dz - B(t, r, s) = 0$$
$$(5.1.5)$$

The initial value requires that

$$U(0, r, s) = f(r, s)$$

Equation (5.1.5) may be interpreted as a linear abstract Cauchy problem for the function $U: [0, T] \to L_2[a, b] \times L_2[a, b]$. We shall see that linear Cauchy problems have global solutions so that $U(t, \cdot, \cdot)$ exists on $[0, T]$ for all $T > 0$. The boundary value problem (5.1.3) will have a solution provided we can find a function $y \in L_2[a, b]$ which satisfies

$$g\left(\int_a^b U(t, r, s) y^2(s) \, ds, y(r) \right) = 0$$

In general, little can be said about solutions of such equations without specific information on U and g. We note in passing that if $g(x, y) = x - y + a$, then the boundary condition leads to the Hammerstein integral equation

$$\int_a^b U(T, r, s) y^2(s) \, ds + a(r) = y(r)$$

Such equations are discussed in Vainberg (1964).

If the boundary value problem (5.1.3) is to be solved numerically, one can discretize the interval $[a, b]$, replace the integrals by sums and, possibly, apply invariant imbedding to the resulting finite-dimensional system. It is to be expected that a representation analogous to (5.14) will apply to the reduced equations. On the other hand, one may attack Eq. (5.1.5) directly, for example with a Picard iteration applied to the equivalent formulation

$$U(t, r, s) = f(r, s) + \int_0^t \left[-2D(\tau, r, s)U(\tau, r, s) \right.$$
$$\left. + \int_a^b A(\tau, r, z)U(\tau, z, s) \, dz + B(\tau, r, s) \right] d\tau$$

As a consequence, we see that there exist two alternative approaches to solving abstract boundary value problems numerically by invariant imbedding. Either the infinite-dimensional problem is reduced to finite dimensions and the resulting system is solved by invariant imbedding, or the abstract invariant imbedding equation is reduced to a numerically manageable form. It likely will depend on the structure of the given problem, which approach is easier to use numerically and to justify theoretically. Of course, if the invariant imbedding equation can be solved in the appropriate infinite dimensional space, even if only approximately, many of the usually thorny convergence proofs can be omitted which accompany the finite-dimensional approximation approach (N.1.2).

In summary we have shown that invariant imbedding applies alike to finite- and infinite-dimensional boundary value problems for differential equations with a smooth right-hand side. It also appears possible to extend the invariant imbedding approach to differential equations with nonsmooth right-hand sides such as nonlinear (differential) evolution equations (roughly the abstract formulation of nonlinear partial differential equations with real characteristics). Whether this approach is useful for abstract problems largely depends on how tractable is the invariant imbedding equation. In general we do not hold much hope for the applicability of characteristic theory to such equations. Indeed, the succeeding sections will show that even linear equations with nonsmooth right-hand sides already produce sizeable complications (see N.1.3).

5.2. INVARIANT IMBEDDING FOR LINEAR EVOLUTION EQUATIONS

The simplifications observed for finite-dimensional linear problems also apply to abstract linear boundary value problems. Moreover, we no longer have to restrict ourselves to differential equations with differentiable right-hand sides; rather we can work with general evolution equations on arbitrary

Banach spaces. Let us first collect some background material on evolution equations. More detailed results may be found in the papers cited and in the books of Butzer and Berens (1967), Dunford and Schwartz (1957), Friedman (1969), Hille and Phillips (1957), and Kato (1966).

DEFINITION 5.2.1. Let $\{A(t)\}$ be a family of linear operators from a subspace $D(A(t)) \subset X$ into X defined for $t \in (0, T)$, $T > 0$. Then the equation

$$x' = A(t)x + F(t), \qquad x(0) = x_0 \tag{5.2.1}$$

is called a linear *evolution equation*.

Throughout this discussion we shall assume that the domain of $A(t)$ is independent of t. If $A(t)$ is bounded on X this will always be the case. The domain of $A(t)$ will be denoted by $D(A)$.

Associated with the linear operator $A(t)$ is the fundamental solution of the homogeneous equation

$$x' = A(t)x, \qquad x(0) = x_0 \tag{5.2.2}$$

Because A is not necessarily bounded, this fundamental solution may have continuity properties which do not have a direct analog in a finite-dimensional setting. We shall use the following terms.

DEFINITION 5.2.2. An abstract function $\phi: [0, T] \to L(X, Y)$ is

(i) *uniformly continuous* if $\phi(t)$ is continuous in t in the topology of $L(X, Y)$ and

(ii) *strongly continuous* if $\phi(t)x$ is continuous in t for each x in the topology of Y.

We note that for finite-dimensional spaces X, Y continuity (i) is the usual matrix continuity. We now can introduce the fundamental solution for (5.2.2) which we shall call an evolution operator.

DEFINITION 5.2.3. A family $\{\phi(t, s)\}$ of bounded linear operators on X defined for $0 \leq s \leq t \leq T$ and satisfying the following conditions:

(i) $\phi(t, s)$ is strongly continuous in both variables,
(ii) $\phi(t, s)\phi(s, r) = \phi(t, r)$, $r \leq s \leq t$,
(iii) $(\partial\phi/\partial t)(t, s)x_0 = A(t)\phi(t, s)x_0$,
(iv) $(\partial\phi/\partial s)(t, s)x_0 = -\phi(t, s)A(s)x_0$

for $x_0 \in D(A)$ is called an *evolution operator* with generator $A(t)$.

Let us look at some examples of evolution operators and their generators. If $A(t)$ is an $n \times n$ matrix then $\phi(t, r)$ is the usual fundamental matrix of the system $\phi'(t, r) = A(t)\phi(t, r)$, $\phi(r, r) = I$. It is well known that the fundamental matrix has the stronger group property

$$\phi(t, r)\phi(r, s) = \phi(t, s) \qquad \text{for} \quad t, s, r \in [0, T]$$

rather than only the so-called evolution property (ii). Considering now infinite-dimensional spaces, we know that the evolution operator exists and has the group property whenever $A(t) \in L(X, X)$, and in this case (iii) and (iv) hold for all $x_0 \in X$. This result follows from the usual coordinate free presentation of differential equations (Dieudonné, 1969). In addition, if $A(t)$ is continuous in t then $\phi(t, s)$ is uniformly continuous in t and s. A more complicated class of evolution operators is obtained if A is time-independent but unbounded. A typical candidate in this context is the operator A defined by

$$(Ax)(s) = \partial x(s)/\partial s + \int_0^\infty e^{-rs}x(r) \, dr$$

for all differentiable functions $x(s)$ (Dunford and Schwartz, 1957, p. 656). Not all unbounded operators generate evolution operators, but whenever an evolution operator $\phi(t, r)$ can be paired with a time-independent unbounded operator A it is called a semigroup and Definition 5.2.3 can be specialized to:

DEFINITION 5.2.4. A family $\{T(t)\}$ of bounded linear operators on X defined for $0 \le t < \infty$ and satisfying the following conditions

(i) $T(t)$ is strongly continuous in t
(ii) $T(s + t) = T(s)T(t)$, $s, t \ge 0$, $T(0) = I$

is called a *one parameter semigroup*.

DEFINITION 5.2.5. For a given semigroup $\{T(t)\}$ and $h > 0$ set

$$A_h x = ((T(h) - I)/h)x, \qquad x \in X$$

and define the operator A by

$$Ax = \lim_{h \to 0} A_h x$$

Let $D(A)$ be the set of all x for which this limit exist, then A is the *infinitesimal generator* of the semigroup $\{T(t)\}$ and $D(A)$ is its domain.

It is a consequence of this definition that A is a closed linear operator and that it satisfies conditions (iii) and (iv) in Definition 5.2.3. In other words, we have the following relationship between $\{T(t)\}$ and its generator A

$$\frac{\partial T}{\partial t}(t - r)x = AT(t - r)x, \qquad \frac{\partial T}{\partial t}(t - r)x = -T(t - r)Ax$$

for all $x \in D(A)$. Thus, in terms of the evolution operator ϕ we have

$$\phi(t, r) = T(t - r)$$

Semigroups, their properties and application have been studied extensively in the literature cited, where emphasis is usually placed on classifying those operators which generate semigroups and on examining the properties of the resulting semigroup.

The next level of complication would be the use of time-dependent generators $A(t)$ as discussed in Kato (1964) and leading to strongly continuous evolution operators. The theory now becomes complicated in detail; we shall not discuss this more general case.

Given an evolution operator $\phi(t, s)$ corresponding to the linear operator $A(t)$, the inhomogeneous equation (5.2.1) formally has the variation of constants solution

$$x(t) = \phi(t, 0)x_0 + \int_0^t \phi(t, s)F(s)\, ds \tag{5.2.3}$$

Again, if $A(t)$ and $F(t)$ are continuous in t with values in $L(X, X)$ and X, respectively, the function $x(t)$ is differentiable in t. Classical (differentiable) solutions of (5.2.1) are generally called *strong solutions*. On the other hand, the Volterra integral equation (5.2.3) may have a continuous solution $x(t)$ which is not necessarily differentiable. In this case $x(t)$ is known as a *mild solution* of (5.2.1). The integral of (5.2.3) may now be interpreted as a Bochner integral (Hille and Phillips, 1957, p. 78). If X is finite-dimensional, then the mild solution of (5.2.1) is called a *Caratheodory solution*. For the remainder of this section we shall deal exclusively with strong solutions.

Let us now turn to the characteristic theory for linear evolution equations. In particular, differential equations and initial values of the following type will be considered.

$$\begin{aligned} x' &= A(t)x + B(t)y + F(t), & x(0) &= fs + a \\ y' &= C(t)x + D(t)y + G(t), & y(0) &= s \end{aligned} \tag{5.2.4}$$

In this formulation x and y are abstract functions defined on $[0, T]$ with values in some arbitrary Banach spaces X and Y, respectively. For fixed t the operators A, B, C, and D are linear (but not necessarily bounded) operators with domain yet to be specified, while the source terms F and G again are abstract functions; f is also a linear operator and a and s belong to given subsets of X and Y, respectively.

Under specific conditions on the functions in (5.2.4) which assure the existence of solutions for (5.2.4) and which allow elimination of the parameters, we can obtain a characteristic theory analogous to that of Section 1.2. The basic result again is the existence of an affine mapping relating $x(t)$ and $y(t)$. However, rather than the constructive approach found in the discussion leading to Theorem 1.2.1, the converse is of interest and use.

THEOREM 5.2.1. Suppose that the initial value problem

$$U' = B(t) + A(t)U - UD(t) - UC(t)U, \qquad U(0) = f \qquad (5.2.5a)$$

$$w' = [A(t) - U(t)C(t)]w - U(t)G(t) + F(t), \quad w(0) = a \qquad (5.2.5b)$$

$$y' = [C(t)U(t) + D(t)]y + C(t)w(t) + G(t), \quad y(0) = s \qquad (5.2.5c)$$

has a solution $\{U(t), w(t), y(t)\}$ such that $U(t)y(t)$, $w(t)$ and $y(t)$ are strongly differentiable on $(0, T)$, then $\{x(t) \equiv U(t)y(t) + w(t), y(t)\}$ is a solution of the initial value problem (5.2.4).

The proof follows by straightforward differentiation and substitution of the defining Eqs. (5.2.5a–c).

COROLLARY 5.2.1. If Theorem 5.2.1 applies with $y(0) = s$ replaced by $y(T) = b$, then $\{x(t) \equiv U(t)y(t) + w(t), y(t)\}$ is a solution of the boundary value problem

$$
\begin{aligned}
x' &= Z(t)x + B(t)y + F(t), & x(0) &= fy(0) + a \\
y' &= C(t)x + D(t)y + G(t), & y(T) &= b
\end{aligned}
\qquad (5.2.6)
$$

The proof again follows by verification that x and y satisfy Eqs. (5.2.6). In addition, if the initial value problem (5.2.4) has a unique solution, then Eqs. (5.2.5) necessarily must have unique solutions. Moreover, if (5.2.4) has a unique solution for arbitrary s in a given set D, and if the solution, $y(0)$ of (5.2.6) belongs to D, then the solution of the boundary value problem (5.2.6) is unique as well. In this case the solution of (5.2.6) necessarily is imbedded into the integral surface $x(t, y) = U(t)y + w(t)$ generated with the shooting method. Note that $U(t)$ for fixed t is a linear operator from Y to X. In all applications below $U(t)$ actually belongs to $L(Y, X)$.

In summary, if Corollary 5.2.1 applies, the two-point boundary value problem (5.2.6) can be solved by integrating in succession the initial value problems (5.2.5a–c). A comparison with the results of Section 2.6 shows that Eqs. (5.2.5a and b) are exactly the invariant imbedding equations corresponding to the boundary value problem (5.2.6).

The difficulty in applying Corollary 5.2.1 lies in establishing the existence of differentiable solutions for the Riccati equation. In Section 5.3 it is shown that we need only mild solutions for Eqs. (5.2.5), while the existence of mild solutions for abstract Riccati equations is discussed in Chapter 6. Here we shall cite some examples where a strongly differentiable solution can be generated, at least over small intervals, with the method of characteristics.

Example 5.2.1. Radiative Transfer in an Isotropic Slab

Steady-state radiative transfer in a plane-parallel slab furnishes a simple but nontrivial abstract boundary value problem involving bounded linear operators. It is well known [see, e.g., Bellman *et al.* (1963) or Wing (1962)] that the following integral equation describes the particle density within the slab

$$\mu(\partial N/\partial z)(z, \mu) + \sigma N(z, \mu) = (c/2) \int_{-1}^{1} N(z, \lambda) \, d\lambda$$

$$N(0, \mu) = a(\mu), \qquad \mu > 0 \qquad \qquad (5.2.7)$$

$$N(Z, \mu) = b(\mu), \qquad \mu > 0$$

where z and μ denote the spacial and angular dependence of the density. We shall assume for simplicity that σ and c are constants, although the development below will remain valid when $\sigma = \sigma(z, \mu)$ and $c = c(z, \mu)$.

The system (5.2.7) is clearly subject to two-point boundary conditions but needs reinterpretation before it can be cast into the form (5.2.6). First of all we shall assume that μ is restricted to the intervals $[-1, -\varepsilon]$ and $[\varepsilon, 1]$ for some small $\varepsilon > 0$ to remove the singularity at the origin. We set

$$x(z, \mu) = N(z, \mu), \qquad \mu \in [\varepsilon, 1]$$

$$y(z, \mu) = N(z, -\mu), \qquad \mu \in [\varepsilon, 1]$$

Then equation (5.2.7) can be rewritten as

$$(\partial x/\partial z)(z, \mu) + (\sigma/\mu)x(z, \mu) = (c/2\mu) \int_{\varepsilon}^{1} [x(z, \lambda) + y(z, \lambda)] \, d\lambda$$

$$(5.2.8)$$

$$(\partial y/\partial z)(z, \mu) - (\sigma/\mu)y(z, \mu) = -(c/2\mu) \int_{\varepsilon}^{1} [x(z, \lambda) + y(z, \lambda)] \, d\lambda$$

where $\mu \in [\varepsilon, 1]$. Suppose that we are searching for solutions x and y which are differentiable in z and continuous in μ so that $x(z, \cdot)$ and $y(z, \cdot)$ belong to the space $C[\varepsilon, 1]$ of continuous functions on $[\varepsilon, 1]$. In analogy to the example of Section 5.1 we can rewrite the above boundary problem as

$$x' = Ax + By, \qquad x(0) = a$$
$$y' = -Bx - Ay, \qquad y(Z) = b \qquad (5.2.9)$$

where we assume that a and b are continuous in μ, and where A and B are the bounded linear operators in L $(C[\varepsilon, 1], C[\varepsilon, 1])$ defined by

$$Ah = (c/2\mu) \int_{\varepsilon}^{1} h(\lambda)\, d\lambda - (\sigma/\mu)h(\mu)$$

$$Bh = (c/2\mu) \int_{\varepsilon}^{1} h(\lambda)\, d\lambda$$

for arbitrary $h \in C[\varepsilon, 1]$.

Invariant imbedding states that the solution of (5.2.9) can be expressed as

$$x(z) = U(z)y(z) + w(z)$$

where $U \colon [0, b] \to L(C[\varepsilon, 1], C[\varepsilon, 1])$ and $w \in C[\varepsilon, 1]$ are solutions of

$$U' = B + AU + UA + UBU, \qquad U(0) = 0 \qquad (5.2.10)$$
$$w' = [U(z)B + A]w, \qquad\qquad w(0) = a \qquad (5.2.11)$$

Equation (5.2.10) is an abstract differential equation with bounded right-hand side on the Banach space $L(C[\varepsilon, 1], C[\varepsilon, 1])$ and always has a local solution. In fact, a quantitative estimate can be obtained from Theorem 5.1.3 after estimating the size of A and B. We note

$$\| A \| = \sup_{\|x\|=1} \| Ax \| = \sup_{\|x\|=1} \left\{ \max_{\mu \in [\varepsilon, 1]} \left| (c/2\mu) \int_{\varepsilon}^{1} x(\lambda)\, d\lambda - (\sigma/\mu)x(\mu) \right| \right\}$$

$$\| A \| \leq K/\varepsilon \quad \text{for some constant } K > 0$$

Similarly, we find that $\| B \| \leq K/\varepsilon$ provided K is chosen sufficiently large. Hence by Theorem 5.1.3 the boundary value problem (5.2.9) has a unique solution provided

$$Z < (\varepsilon/3K) \ln(4)$$

Since in general $\varepsilon \to 0$, this result is little illuminating. We shall remedy the situation in Chapter 6.

The utility of the invariant imbedding formulation for the Boltzmann equation (5.2.7) is due to the fact that a concrete representation can be found for the operator U of (5.2.10). It follows from a theorem of Dunford and Schwartz (1957, p. 490) that for fixed z there exists a mapping τ from $[\varepsilon, 1]$ into the dual space of $C[\varepsilon, 1]$ such that for arbitrary $h \in C[\varepsilon, 1]$

$$(Uh)(\mu) = \tau(\mu)h$$

The Riesz representation theorem for linear functionals on $C[\varepsilon, 1]$ assures that $\tau(\mu)h$ can be written as the Stieltjes integral

$$\tau(\mu)h = \int_\varepsilon^1 h(\lambda)u(\mu, d\lambda)$$

where $u(\mu, \lambda)$ is continuous in μ and of bounded variation in λ. Let us go one step further and *assume* that the solution U of (5.2.10) can actually be expressed as the Riemann integral

$$(U(z)h)(\mu) = \int_\varepsilon^1 h(\lambda)u(z, \mu, \lambda) \, d\lambda \tag{5.2.12}$$

Substitution of (5.2.12) into (5.2.10) yields

$$\int_\varepsilon^1 \left[(\partial u/\partial z)(z, \mu, \lambda) - (c/2\mu) + (\partial/\mu + \partial/\lambda)u(z, \mu, \lambda) \right.$$
$$- (c/2\mu) \int_\varepsilon^1 u(z, \eta, \lambda) \, d\eta - \int_\varepsilon^1 u(z, \mu, \eta)(c/2\eta) \, d\eta$$
$$\left. - \int_\varepsilon^1 \int_\varepsilon^1 u(z, \mu, \varrho)(c/2\varrho)u(z, \eta, \lambda) \, d\varrho \, d\eta \right] h(\lambda) \, d\lambda = 0$$

Since this equation has to hold for all continuous functions h the integrand has to vanish and we obtain the reduced invariant imbedding equation

$$(\partial u/\partial z)(z, \mu, \lambda) - c/2\mu + (\sigma/\mu + \sigma/\lambda)u(z, \mu, \lambda) - (c/2\mu) \int_\varepsilon^1 u(z, \eta, \lambda) \, d\eta$$
$$- \int_\varepsilon^1 u(z, \mu, \eta)(c/2\eta) \, d\eta - \int_\varepsilon^1 \int_\varepsilon^1 u(z, \mu, \varrho)(c/2\varrho)u(z, \eta, \lambda) \, d\varrho \, d\eta = 0 \tag{5.2.13}$$

In order to derive (5.2.13), we had to assume the validity of the representation (5.2.12). However, if an independent existence proof for (5.2.13) is available, we can make all arguments rigorous because Corollary 5.2.1

applies. Suppose, for example, that (5.2.13) has a solution $u(z, \mu, \lambda)$ on $[0, z] \times [\epsilon, 1] \times [\epsilon, 1]$ which is differentiable in z and continuous in μ and λ. Then the operator $U(z)$ defined on $C(\epsilon, 1]$ by (5.2.12) is necessarily the unique solution of (5.2.5a). Furthermore, if $a(\mu) \equiv 0$, then the solution w of the linear equation (5.2.1) satisfies $w(z) \equiv 0$ and hence

$$N(Z, \mu) = \int_{\epsilon}^{1} u(Z, \mu, \lambda) b(\lambda) \, d\lambda$$

is the correct boundary value for the Boltzmann equation (5.2.7). If $a(\mu) \not\equiv 0$, then it remains to integrate the linear equation (5.2.5b) which now assumes the form

$$(\partial w / \partial z)(z, \mu) = (c/2\mu) \int_{\epsilon}^{1} w(z, \lambda) \, d\lambda - (\sigma/\mu) w(z, \mu)$$

$$+ \int_{\epsilon}^{1} \int_{\epsilon}^{1} u(z, \mu, \lambda)(c/2\lambda) w(z, \eta) \, d\eta \, d\lambda, \qquad w(0, \mu) = a(\mu)$$

A linear equation with a bounded right-hand side will always have a unique solution. Hence $w(z, \mu)$ exists. This in turn allows us to integrate the linear equation (5.2.5c) to find $y(z, \mu)$. Since this equation also has a unique solution, Corollary 5.2.1 applies to this problem.

It should be pointed out that the abstract equations (5.2.9) corresponding to (5.2.7) may just as well be interpreted in the Banach space $L_p(\epsilon, 1)$, $p \geq 1$ (N.2.1). Of particular interest is the L_∞ setting because for this case we do have an independent existence proof for (5.2.13). Indeed, if we set

$$u(z, \mu, \lambda) = (2\pi\lambda/\mu) R_B(z, \mu, -\lambda)$$

and substitute u into Eq. (5.2.1), we obtain exactly the invariant imbedding equation for R_B derived earlier by Bailey (1964) from a boundary perturbation point of view. Moreover, Bailey has shown that R_B exists as a bounded measurable function on $[0, Z] \times [0, 1] \times [-1, 0]$ provided

$$(c/\sigma) \int_{0}^{1} (1 - e^{-\sigma Z/\lambda}) \, d\lambda < 1$$

This, of course, is a useful result compared to the local estimate given above. Finally, we would like to mention that the invariant imbedding equation of Wing (1962) for transport in a slab is recovered when we substitute the expression

$$u(z, \mu, \lambda) = (1/2\pi) R_W(2, \mu, -\lambda)$$

into (5.2.13) where R_W is the reflection function of Wing. ∎

Example 5.2.1 contains all the features commonly present when invariant imbedding is applied to linear two-point boundary value problems for ordinary, partial, or integro-differential equations. First, the two differential equations for x and y in (5.2.4) are identified, either in an abstract setting (like (5.2.9)) or in concrete form (like (5.2.8)). We then can formally write down the invariant imbedding equations (5.2.10) and (5.2.11). Next, a representation is guessed for the operator U which on substitution into the Riccati equation leads to the final concrete invariant imbedding equation. And, if this invariant imbedding equation can be shown to have a solution, then Corollary 5.2.1 assures that we have indeed solved the original two-point boundary value problem. It may be observed how each of these steps recurs in the next two examples.

If the linear operators in (5.2.6) no longer are bounded but generate bounded evolution operators with the group property, it still is quite easy to verify that the Eqs. (5.2.5) have strong solutions because the solution U of (5.2.5a) can again be constructed exactly as outlined in Section 2.6. Let us illustrate this assertion with an example.

Example 5.2.2. Time-Dependent One-Dimensional Transport

According to Wing (1962) neutron transport in a rod is described by the hyperbolic system

$$\left(\frac{\partial u}{\partial z} + \frac{1}{c}\frac{\partial u}{\partial t}\right)u(z, t) = \sigma(F - 1)u(z, t) + \sigma Bv(z, t)$$
$$\left(-\frac{\partial v}{\partial z} + \frac{1}{c}\frac{\partial v}{\partial t}\right)v(z, t) = \sigma Bu(z, t) + \sigma(F - 1)v(z, t)$$
(5.2.14)

where u and v represent particle fluxes to the right and left within the rod. We shall assume that all coefficients are constants. Moreover, for ease of comparison with the results of Bailey (1964) we shall suppose that $B = F = 1$ so that the transport equations are given as

$$\left(\frac{\partial}{\partial z} + \frac{1}{c}\frac{\partial}{\partial t}\right)u(z, t) = \sigma v(z, t)$$

$$\left(-\frac{\partial}{\partial z} + \frac{1}{c}\frac{\partial}{\partial t}\right)v(z, t) = \sigma u(z, t)$$
(5.2.15)

$$u(0, t) = a(t), \qquad v(Z, t) = b(t)$$

With $A = (1/c)(\partial/\partial t)$ the system (5.2.15) can formally be rewritten as

$$u'(z) = -Au + \sigma u, \qquad v'(z) = -\sigma u + Av$$
(5.2.16)

where u and v are to be interpreted as elements in a function space yet to be defined. The corresponding invariant imbedding equations are the Riccati equation

$$U' = \sigma 1 - AU - UA + \sigma UIU, \qquad u(0) = 0 \qquad (5.2.17)$$

and the linear equation

$$w' = [\sigma U(z) - A]w, \qquad w(0) = a$$

If the data functions in (5.2.15) are integrable on $(-\infty, \infty)$, then the abstract system (5.2.16) can be interpreted as an evolution equation on the space $X = L(-\infty, \infty) \times L(-\infty, \infty)$, where $L(-\infty, \infty)$ is the Banach space of integrable functions on $(-\infty, \infty)$ with the usual L_1 norm, where X is normed by $\| (x, y) \|_X = \max\{\| x \|_{L_1}, \| y \|_{L_1}\}$, and where A is defined only for elements in the subspace D of absolutely continuous functions.

It is well known that A is the closed linear generator of the strongly continuous translation group (evolution operator) $\{T(z)\}$ defined on $[0, Z]$ [see, e.g., Dunford and Schwartz (1957, p. 630)]. Moreover, bounded perturbations of the generator of a group again are group generators. Hence it follows that Eqs. (5.2.16) subject to initial values in D have a unique strong solution. Integrating (5.2.16) subject to $v(0) = s \in D$ and using the invertibility of the evolution operator for (5.7.16) we can proceed exactly as in Section 2.6 and construct the affine transformation

$$u = U(z)v + w(z)$$

where $U(z)s$ is a solution of (5.2.17) for arbitrary $s \in D$. As in the case of transport in a slab, one may guess a concrete representation for the linear operator $U(z)$. Guided by the discussion of Duhamel's integral (Courant and Hilbert, 1962, pp. 512–513), one usually assumes that $U(z)v$ for arbitrary $v \in L(-\infty, \infty)$ can be written as the convolution

$$U(z)v(t) = \int_{-\infty}^{\infty} R(z, t - \mu)v(\mu) \, d\mu$$

or in short

$$U(z)v(t) = (R \star v)(z, t) \qquad (5.2.18)$$

Substitution of the representation (5.2.18) into the Riccati equation (5.2.17) leads to

$$(\partial/\partial z)(R \star s) = -(1/c)(\partial/\partial t)(R \star s) - (1/c)(R \star \partial s/\partial \mu) + \sigma(s + R \star R \star s)$$
$$R(0, \cdot) \star s = 0 \qquad (5.2.19)$$

for all $s \in D$. This expression can be simplified if we assume that $R(z, \mu)$ is absolutely continuous in μ. Integrating $(R \star \partial s/\partial \mu)$ by parts and setting $s = \delta \star s$ where δ is the Dirac delta function we obtain the equation

$$[\partial R/\partial z + (2/c)(\partial R/\partial t) - \sigma R \star R - \sigma\delta] \star s = 0$$

$$R(0, \cdot) \star s = 0$$

This equation has to hold for all functions s with $s' \in L(-\infty, \infty)$; since D is dense in $L(-\infty, \infty)$, it follows that R is a solution of the equation

$$(\partial R/\partial z)(z, t - r) + (2/c)(\partial R/\partial t)(z, t - r)$$

$$- \sigma \int_{-\infty}^{\infty} R(z, t - s)R(z, s - r) \, ds - \sigma\delta(t - r) = 0$$

$$\tag{5.2.20}$$

$$R(0, t - r) = 0$$

for $t, r \in (-\infty, \infty)$. Equation (5.2.20) is a concrete representation of the Riccati equation (5.2.17). If $R(z, t)$ exists, it defines a bounded linear operator from $L(-\infty, \infty)$ into itself. Anticipating the results of the next section, we can state that $[\sigma U(z) - A]$ generates an evolution operator so that the solution w of (5.2.17) exists. Note that in concrete form w satisfies the integro–differential equation

$$(\partial w/\partial z)(z, t) = \sigma \int_{-\infty}^{\infty} R(z, t - r)w(z, r) \, dr - (\partial w/\partial t)w(z, t)$$

$$w(0, t) = u(t)$$

The proper boundary value for (5.2.15) at $z = Z$ is then given as

$$u(Z, t) = \int_{-\infty}^{\infty} R(Z, t - r)b(r) \, dr + w(Z, t)$$

To conclude our discussion of time dependent transport in a rod, let us show how the invariant imbedding equation of Bailey (1964) can be recovered from (5.2.20). Like Bailey, let us assume that the system is at rest for $t \leq 0$ and that the boundary values are given as

$$u(0, t) \equiv 0, \quad v(Z, t) = b(t) = 1/c, \quad t \in [0, T], \quad b(t) \equiv 0 \quad \text{otherwise}$$

As a first consequence, we see that $R(z, t) = 0$ for $t \leq 0$ and that $w(z, t) \equiv 0$ for all t. Let us define

$$R_B(z, t) = \int_{-\infty}^{\infty} R(z, t - r)b(r) \, dr$$

$R(z, t) = b(t) = 0$ for $t \leq 0$ implies that

$$R_B(z, t) = (1/c) \int_0^t R(z, t - r)\, dr$$

Multiplying (5.2.20) by $b(t)$ and integrating over $[0, t]$, we obtain

$$\frac{\partial R_B}{\partial z}(z, t) + \frac{2}{c} \frac{\partial R_B}{\partial t}(z, t) - \sigma \int_0^t R(z, t - s) R_B(z, s)\, ds - \frac{\sigma}{c} = 0$$
$$R_B(0, t) = 0, \qquad t > 0 \tag{5.2.21}$$

Next we observe that

$$\frac{\partial R_B}{\partial t}(z, t) = \frac{1}{c} R(z, t)$$

Substitution into Eq. (5.2.21) yields the final result

$$\partial R_B / \partial z + (2/c)(\partial R_B / \partial t) - \sigma c(R_B \star \partial R_B / \partial t) - \sigma / c = 0$$
$$t < T, \qquad R_B(0, t) = 0 \tag{5.2.22}$$

This is precisely the invariant imbedding equation derived by Bailey. We remark that Bailey also proves the existence of a unique solution for (5.2.15) when $b(t) = 1/c$. The complete system (5.2.14) is treated in Chapter 6. ∎

Example 5.2.3. Steady-State Transport in a Solid Sphere

Without regard for the attendant mathematical problems, let us formally derive the invariant imbedding equation for steady-state transport in a solid sphere. As we shall see, here we deal with the case where the evolution equation involves unbounded operators depending on the independent variable. The Boltzmann formulation for this problem is (Bailey, 1964)

$$\mu\left(\frac{\partial}{\partial r} + \frac{1 - \mu^2}{r} \frac{\partial}{\partial \mu} + \sigma\right) N(r, \mu) = \frac{\gamma \sigma}{2} \int_{-1}^1 N(r, \lambda)\, d\lambda$$
$$N(0, \mu) = N(0, -\mu), \qquad \mu \in [0, 1]$$
$$N(Z, \mu) = g(\mu), \qquad \mu \in [-1, 0)$$

We shall write this equation in two-point boundary form:

$$\frac{\partial x}{\partial r} = \left(-\frac{1 - \mu^2}{\mu r} \frac{\partial}{\partial \mu} - \frac{\sigma}{\mu} + \frac{\gamma \sigma}{2\mu} \int_0^1 \cdot\, d\lambda\right) x + \left(\frac{\gamma \sigma}{2\mu} \int_0^1 \cdot\, d\lambda\right) y$$
$$\frac{\partial y}{\partial r} = \left(-\frac{\gamma \sigma}{2\mu} \int_0^1 \cdot\, d\lambda\right) x + \left(-\frac{1 - \mu^2}{\mu r} \frac{\partial}{\partial \mu} + \frac{\partial}{\partial \mu} + \frac{\sigma}{\mu} - \frac{\gamma \sigma}{2\mu} \int_0^1 \cdot\, d\lambda\right) y$$
$$\tag{5.2.23}$$
$$x(0) = y(0), \qquad y(Z) = g(\mu)$$

where $x(r, \mu) = N(r, \mu)$, $y(r, \mu) = N(r, -\mu)$, $\mu \in [0, 1]$. These two equations are to be interpreted as abstract differential equations where the terms in the parentheses are operators operating on x and y. In particular, by $\int_0^1 \cdot \, d\lambda$ we mean

$$\int_0^1 \cdot \, d\lambda \, x = \int_0^1 x(\lambda) \, d\lambda$$

where x is an element in some function space.

It is readily seen that the system (5.3.23) can formally be written as

$$x' = A(r)x + By, \qquad x(0) = y(0)$$
$$y' = Cx + D(r)y, \qquad y(Z) = g$$

where A, B, C, and D are identified with the corresponding terms in (5.3.23). Let us now look at the invariant imbedding equations (5.2.5) for this problem. The linear equation (5.2.5b) has the solution $w(r, \mu) \equiv 0$ and can be disregarded while for the Riccati equation (5.2.5a) we shall assume the existence of a solution

$$x(r, \mu) = \int_0^1 U(r, \mu, \lambda)y(\lambda) \, d\lambda$$

where U is as smooth as required for subsequent operations. The right-hand side of the Riccati equation becomes

$$Uy' = \int_0^1 \frac{\partial}{\partial r} U(r, \mu, \lambda)y(\lambda) \, d\lambda$$

Let us next look term by term at the right-hand side. We see that

$$By = \frac{\gamma\sigma}{2\mu} \int_0^1 y(\lambda) \, d\lambda = \int_0^1 \frac{\gamma\sigma}{2\mu} y(\lambda) \, d\lambda$$

$$AUy = \left(\int_0^1 - \frac{1-\mu^2}{\mu r} \frac{\partial}{\partial\mu} U(r, \mu, \lambda)y(\lambda) \, d\lambda - \int_0^1 \frac{\sigma}{\mu} U(r, \mu, \lambda)y(\lambda) \, d\lambda \right.$$
$$\left. + \int_0^1 \frac{\gamma\sigma}{2\mu} \int_0^1 U(r, \eta, \lambda)y(\lambda) \, d\lambda \, d\eta \right)$$

$$UDy = \int_0^1 - U(r, \mu, \lambda) \frac{1-\lambda^2}{\lambda r} \frac{\partial}{\partial\lambda} y(\lambda) \, d\lambda + \int_0^1 U(r, \mu, \lambda) \frac{\sigma}{\lambda} y(\lambda) \, d\lambda$$
$$- \int_0^1 U(r, \mu, \eta) \frac{\sigma\gamma}{2\eta} \int_0^1 y(\lambda) \, d\lambda \, d\eta$$

Integration by parts applied to the first term of $UD(r)y$ yields

$$\int_0^1 U(r, \mu, \lambda) \frac{1 - \lambda^2}{\lambda r} \frac{\partial}{\partial \lambda} y(\lambda) \, d\lambda = U(r, \mu, \lambda) \frac{1 - \lambda^2}{\lambda r} y(\lambda) \Big|_0^1$$

$$+ \int_0^1 \frac{1 + \lambda^2}{\lambda^2 r} U(r, \mu, \lambda) y(\lambda) \, d\lambda$$

$$- \int \frac{1 - \lambda^2}{\lambda r} \frac{\partial U}{\partial \lambda} (r, \mu, \lambda) y(\lambda) \, d\lambda$$

Finally, the quadratic term can be written as

$$UCUy = \int_0^1 U(r, \mu, \xi) \int_0^1 \left(-(\gamma \sigma / 2\xi) \int_0^1 U(r, \eta, \lambda) y(\lambda) \, d\lambda \right) d\eta \, d\xi$$

We shall change, where necessary, the order of integration so that we integrate last with respect to λ. We also shall assume that $\lim_{\lambda \to 0} U(r, \mu, \lambda) \{(1 - \lambda^2)/\lambda r\} y(\lambda) = 0$. Finally, we shall keep in mind that the Riccati equation has to hold for all differentiable y. These postulates allow us to isolate the integrand with respect to λ in the Riccati equation as

$$\frac{\partial u}{\partial r} (r, \mu, \lambda) = \frac{\gamma \sigma}{2\mu} - \frac{1 - \mu^2}{\mu r} \frac{\partial u}{\partial \mu} (r, \mu, \lambda) - \left(\frac{\sigma}{\mu} + \frac{\sigma}{\lambda} \right) U(r, \mu, \lambda)$$

$$+ \frac{\gamma \sigma}{2\mu} \int_0^1 U(r, \eta, \lambda) \, d\eta + \int_0^1 U(r, \mu, \eta) \frac{\sigma \gamma}{2\eta} \, d\eta$$

$$+ \frac{1 + \lambda^2}{\lambda r} U(r, \mu, \lambda) - \frac{1 - \lambda^2}{\lambda r} \frac{\partial u}{\partial \lambda} (r, \mu, \lambda)$$

$$+ \int_0^1 U(r, \mu, \varrho) \frac{\gamma \sigma}{2\varrho} \int_0^1 U(r, \eta, \lambda) \, dy$$

while the initial condition $x(0) = U(0)y$ requires that

$$U(0, \mu, \lambda) = \delta(\mu - \lambda)$$

This equation is precisely the invariant imbedding equation derived by Bailey and Wing (1964). Moreover, a related expression was studied by Bailey (1964) with a view to providing existence and uniqueness theorems. ∎

Our intention here was to demonstrate that the characteristic theory approach furnishes a mechanical means for deriving invariant imbedding equations. It is apparent from the two-point formulation (5.2.23) that a construction of the integral surface $U(r)y$ with the characteristics is a formidable mathematical task because of the complexity of the operators A and D. In the terminology of Definition 5.2.3, A and D would be unbounded time-dependent operators with a singularity at the origin.

Finally, let us briefly comment on the practical use of the invariant imbedding equations (5.2.3), (5.2.20), and (5.2.24). If it is known from physical considerations that the transport system is subcritical, then the Riccati equation will likely have a bounded solution. Moreover, if only the boundary value $x(Z)$ is desired, as is usually the case in neutron transport applications, then one can multiply the invariant imbedding equation with the prescribed boundary value $y(Z)$ and integrate over λ. This simplifies the imbedding equation (compare, for example, (5.2.20) and (5.2.21)). The resulting integro–differential equation can then be attacked numerically with difference and quadrature methods. For the slab equation this approach was used successfully to generate a complete volume of numerical values for the transmission function [see Bellman *et al.* (1963)].

5.3. INVARIANT IMBEDDING FOR MILD SOLUTIONS

The invariant imbedding approach for abstract boundary value problems presented so far always contained a built-in, albeit local, existence and uniqueness proof because the integral surface was generated from the characteristic equations as outlined in Theorem 1.1.2. On the other hand, Theorem 5.2.1 is independent of any concept of characteristics. An alternate approach to proving the existence of solutions for abstract linear boundary value problems is available provided we can establish the existence of solutions for Eqs. (5.2.5a–c). For general evolution equations involving unbounded operators, this is a difficult problem because of the nonlinear Riccati equation and serious domain restrictions affecting the differentiability of such solutions. The problem is simplified if we abandon strong solutions and instead concentrate on mild solutions for boundary value problems as defined below. It may be observed that even in a finite dimensional setting there exists considerable motivation for using mild solutions. One of the common sources of boundary value problems is the theory of optimal control. In many applications the optimal control function may be assumed to be only measurable instead of piecewise continuous as required in earlier examples. This more general framework forces us to deal with Caratheodory solutions. The presentation here is geared to Banach space valued evolution equations with unbounded right-hand sides. Stronger results can be obtained for finite-dimensional systems. For ease of exposition we shall consider only time-independent closed operators leading to semigroups rather than the more general time-dependent generators of evolution operators. The same development is expected to be valid in the latter case (N.3.1). Specifically, throughout this section we shall deal with boundary value problems of the form

$$x' = Ax + B(t)y + F(t), \qquad x(0) = fy(0) + a$$
$$y' = C(t)x + Dy + G(t), \qquad y(T) = gx(T) + b \tag{5.3.1}$$

We shall assume first that B, C, F, and G are continuous abstract functions on $[0, T]$ with values in $L(Y, X)$, $L(X, Y)$, X, and Y, respectively, that f and g are bounded operators on Y and X and that a and b are arbitrary elements in X and Y. As stated, we also shall make the greatly simplifying assumption that

A is the infinitesimal generator of a semigroup $\{V(t),\ 0 \le t \le T\}$;

D is the infinitesimal generator of a semigroup $\{W(t),\ -T \le t \le 0\}$.

We stipulate that the semigroup $\{W(t)\}$ is defined for negative time because in contrast to initial value problems this is the property needed for boundary value problems. It turns out, at least in control problems, that this is a natural postulate, because A and D differ in sign only so that we can define $W(t) = V(-t)$, $t \le 0$.

A mild solution of (5.3.1) is a pair $\{x(t), y(t)\}$ of strongly continuous functions defined on $[0, T]$ which satisfy the integral equations

$$x(t) = V(t)[fy(0) + a] + \int_0^t V(t - r)[B(r)y(r) + F(r)]\, dr$$

$$y(t) = W(t - T)[gx(T) + b] - \int_t^T W(t - r)[C(r)x(r) + G(r)]\, dr \tag{5.3.2}$$

It is easy to verify that $\{x(t), y(t)\}$ is a strong solution of (5.3.1) provided the functions are strongly differentiable.

With the boundary value problem (5.3.1) we can associate the approximation problem depending on the scalar parameter $h > 0$

$$x_h' = A_h x_h + B(t)y_h + F(t), \qquad x_h(0) = fy_h(0) + a$$
$$y_h' = C(t)x_h + D_h y_h + G(t), \qquad y_h(T) = gx_h(T) + b \tag{5.3.3}$$

where A_h and D_h are defined by

$$A_h = (V(h) - 1)/h, \qquad D_h = -(W(-h) - 1)/h \tag{5.3.4}$$

By Definition 5.2.5

$$Ax = \lim_{h \to 0} A_h x, \qquad Dy = \lim_{h \to 0} D_h y$$

for all $x \in D(A)$ and $y \in D(D)$.

Because $V(h)$ and $W(-h)$ are bounded linear operator it follows that A_h and D_h generate uniformly continuous evolution operators

$$V_h \equiv e^{A_h t}, \qquad W_h \equiv e^{D_h (t-T)}$$

invariant imbedding is therefore applicable to (5.3.3). The solution $\{x_h(t), y_h(t)\}$ may be found from

$$U_h' = B(t) + A_h U_h - U_h D_h - U_h C(t) U_h, \qquad U_h(0) = f \qquad (5.3.5a)$$

$$w_h' = [A_h - U_h(t)C(t)]w_h - U_h(t)G(t) + F(t), \qquad w_h(0) = a \qquad (5.3.5b)$$

$$y_h' = [C(t)U_h(t) + D_h]y_h + C(t)w_h(t) + G(t), \qquad (5.3.5c)$$

$$y_h(T) = [1 - gU_h(T)]^{-1}[gw_h(T) + b]$$

$$x_h(t) = U_h(t)y_h(t) + w_h(t)$$

The functions $x_h(t)$ and $y_h(t)$ are continuously differentiable and satisfy (5.3.2) and (5.3.3). In addition, it follows by differentiation that the solutions of (5.3.5) also satisfy the Volterra equations

$$U_h(t) = V_h(t)fW_h(-t) + \int_0^t V_h(t-r)[B(r) - U_h(r)C(r)U_h(r)]W_h(r-t)\,dr$$

$$w_h(t) = R_h(t, 0)a + \int_0^t R_h(t, r)[F(r) - U_h(r)G(r)]\,dr \qquad (5.3.6)$$

$$y_h(t) = S_h(t, T)[1 - gU_h(T)]^{-1}[gw_h(T) + b]$$
$$- \int_T^t S_h(r, t)[C(r)w_h(r) + G(r)]\,dr$$

where R_h and S_h are the uniformly continuous evolution operator generated by $[A_h - U_h(t)C(t)]$ and $[C(t)U_h(t) + D_h]$, respectively.

We are interested in the relationship between (5.3.3) and (5.3.1) as $h \to 0$. A few well-known preliminary results are needed.

LEMMA 5.3.1. $V(t)x = \lim_{h \to 0} V_h(t)x$ uniformly in $t \in [0, T]$ for all $x \in X$.

The proof depends on the observation that $V_h(t)x$ is bounded uniformly with respect to h. Details may be found in Dunford and Schwartz (1957, p. 621).

LEMMA 5.3.2. If A is the generator of a strongly continuous semigroup $\{V(t)\}$ and B is a strongly continuous function on $[0, T]$ with values in $L(X, X)$, then $A + B(t)$ is the generator of a strongly continuous evolution operator $\{\phi(t, r), 0 \leq r \leq t \leq T\}$.

This is a standard perturbation theorem for the generator of a semigroup. A proof may be found in Segal (1963).

Let us now turn to the solution of (5.3.5) and (5.3.6) as $h \to 0$.

LEMMA 5.3.3. The Riccati equation (5.3.5a) has a mild solution for $h = 0$ which satisfies the Volterra equation

$$U(t) = V(t)fW(-t) + \int_0^t V(t - r)[B(r) - U(r)C(r)U(r)]W(r - t)\, dr$$

Proof. Since V and W are strongly continuous, the Picard iteration

$$U^{n+1}(t) = V(t)fW(-t) + \int_0^t V(t - r)[B(r) - U^n(r)C(r)U^n(r)]W(r - t)\, dr$$
$$U^0(t) = V(t)fW(-t)$$

defines a sequence $\{U^n(t)\}$ of strongly continuous functions from $[0, T]$ to $L(Y, X)$. Since V and W are uniformly bounded and measurable (Dunford and Schwartz, 1957) on $[0, T]$, we obtain the estimate

$$\| U^{n+1}(t) \| \leq K_1 + K_2 \int_0^t \| U^n(r) \|^2\, dr$$

for two constants K_1, K_2 and all $n > 0$. It follows from this inequality, that there exists a $\hat{t} > 0$ such that $\| U^n(t) \| < \infty$ uniformly with respect to t and n over compact subsets of $[0, \hat{t}]$. For example, if $\hat{t}$ is chosen such that $K_1 K_2 \hat{t} < \frac{1}{4}$ and $K_2 \| U^0(t) \| \hat{t} < \frac{1}{2}$ for all $t \in [0, \hat{t}]$ it is seen that $K_2 \| U^n(t) \| \hat{t} < \frac{1}{2}$ for all n and $t \in [0, \hat{t}]$. The bound on $\| U^n(t) \|$ is used to show that $\| U^n(t) - U^m(t) \| \to 0$ as $n, m \to \infty$; hence $U^n(t)$ converges pointwise to an operator $U(t) \in L(Y, X)$. Similarly, we see that $\| U^n(t)y - U^m(t)y \| \to 0$ uniformly for arbitrary $y \in Y$. Since each $U^n(t)y$ is uniformly continuous on $[0, \hat{t}]$, it follows that the limit function is also uniformly continuous. Hence, $U(t)$ is strongly continuous. ∎

We remark that the existence of a strongly continuous mild solution $U(t)$ of (5.3.5a) for $h = 0$ and Lemma 5.3.2 assure the existence of strongly continuous evolutions operators

$$S(t, r) \quad \text{with generator} \quad A - U(t)C(t)$$
$$R(t, r) \quad \text{with generator} \quad C(t)U(t) + D$$

Consequently, the solutions $w(t)$ and $y(t)$ of (5.3.5b and c) exist over $[0, \hat{t}]$.

Similar (but simpler) estimates show that $\| U_h(t) \| < \infty$ on $[0, \hat{t}]$ for $h > 0$ so that it becomes a straightforward application of the triangle and Gronwall's inequalities to establish that $\| (U(t) - U_h(t))y \| \to 0$ uniformly in t on $[0, \hat{t}]$ as $h \to 0$ which, in turn, implies that the solutions w and y of (5.3.6) satisfy $\| w(t) - w_h(t) \| \to 0$ and $\| y(t) - y_h(t) \| \to 0$ as $h \to 0$. Consequently, we can state.

THEOREM 5.3.1. Suppose that for all $0 \leq h \leq \hat{h}$ the Volterra equations (5.3.6) have solutions $U_h(t)$, $w_h(t)$, and $y_h(t)$ defined over $[0, T]$. Then the functions $x_h(t) = U_h(t)y_h(t) + w_h(t)$, $y_h(t)$ are solutions of (5.3.3). In particular, for $h = 0$ we obtain a mild solution of (5.3.1).

Proof. For $h > 0$ the functions x_h and y_h are differentiable and satisfy (5.3.3) and hence

$$x_h(t) = V_h(t)[fy_h(0) + a] + \int_0^t V_h(t - r)[B(r)y_h(r) + F(r)]\, dr$$

$$y_h(t) = W_h(t - T)[gx(T) + b] - \int_t^T W_h(t - r)[C(r)x_h(r) + G(r)]\, dr$$

The result follows by letting $h \to 0$. ∎

It was postulated at the beginning of this section that the data functions were continuous in t. By working with mild solutions, we are able to relax this requirement. For example, consider the system

$$\begin{aligned} x_n' &= Ax_n + B_n(t)y_n + F_n(t), & x_n(0) &= fy_n(0) + a \\ y_n' &= C_n(t)x_n + Dy_n + G_n(t), & y_n(T) &= gx_n(T) + b \end{aligned} \tag{5.3.7}$$

and the associated Volterra equations

$$U_n(t) = V(t)fW(-t) + \int_0^t V(t - r)[B_n(r) - U_n C_n(r)U_n]W(r - t)\, dr$$

$$w_n(t) = R(t, 0)a + \int_0^t R(t, r)[F_n(r) - U_n(r)G_n(r)]\, dr$$

$$y_n(t) = S(t, T)[I - gU_n(T)]^{-1}[gw_n(T) + b]$$
$$\qquad - \int_T^t S(r, t)[C_n(r)w_n(r) + G_n(r)]\, dr$$

where the subscripted functions are strongly continuous in t on $[0, T]$ and taken to be an approximation of the corresponding functions in (5.3.1). We readily obtain the following result.

THEOREM 5.3.2. Suppose that $B_n \to B$, $C_n \to C$, $F_n \to F$, and $G_n \to G$ uniformly in t. If the functions U_n, w_n, y_n exist for all $n > N$, and if the solutions U, w, y of (5.3.6) (for $h = 0$) exist over $[0, T]$, then the pair $\{x(t) \equiv U(t)y(t) + w(t), y(t)\}$ is a mild solution of (5.3.1).

Proof. It follows from Theorem 5.3.3 that $x_n(t) \equiv U_n(t)y_n(t) + w_n(t)$, $y_n(t)$ is a mild solution of (5.3.7) for all n. The result follows by letting $n \to \infty$. ∎

Let us summarize these results and provide some examples. First, we have shown that invariant imbedding remains applicable to linear evolution equations which may possess only mild (i.e., strongly continuous but not necessarily differentiable) solutions, provided A and D generate semigroups $\{V(t)\}$, $\{W(-t)\}$, $0 \leq t \leq T$. The basis of this generalization of invariant imbedding is the observation that the Riccati transformation

$$x(t) = U(t)y(t) + w(t)$$

remains valid for the linear system (5.3.1) even when all functions are merely continuous and no longer differentiable. Within the framework of mild solutions we need not require that V and W be, for example, C_0 or holomorphic semigroups; these properties come into play when establishing that mild solutions are also strong solutions. The uniqueness of the mild solution assures that invariant imbedding will yield the correct solution for (5.3.1) provided that this system has a strong or mild solution, and provided that the abstract operator valued function $U(t)$ exists over $[0, T]$ (see N.3.2).

Theorem 5.3.2 or analogous results are useful for treating finite and infinite dimensional systems of differential equations with integrable but not necessarily continuous source terms. For example, suppose that $F: [0, T] \to X$ and $G: [0, T] \to Y$ are Bochner integrable abstract functions. It is known (Hille and Phillips, 1957, p. 86) that continuous functions are dense in the space of Bochner integrable functions. This implies that for given F and G there exist sequences $\{F_n\}$, $\{G_n\}$ of continuous abstract functions such that

$$\int_0^T \| F_n(r) - F(r) \| \, dr \to 0 \quad \text{and} \quad \int_0^T \| G_n(r) - G(r) \| \, dr \to 0$$

as $n \to \infty$. Theorem 5.3.1 applies to system (5.3.1) with F and G replaced by F_n and G_n. A slight modification of the wording of Theorem 5.3.2 allows the conclusion that the mild solution of (5.3.1) can be found from (5.3.6).

Abstract boundary value problems of the general form (5.3.1) arise naturally in optimal control problems when the state equation is a partial differential equation which can be written as an evolution equation. In particular, this approach applies to control problems for parabolic and hyperbolic equations. There exists an extensive literature on this subject. We refer to the monograph of Lions (1971), where the mathematical foundations for an infinite dimensional control theory are developed in great detail under precisely stated conditions. In particular, the role of the infinite dimensional Riccati equation in solving abstract boundary value problems for Hamiltonian systems is explained, although not from the characteristic theory point of view invoked here to help visualize the connection between boundary and initial value problems. Additional theory and numerous applications for controlled partial differential equations, so-called distributed control systems, may be found in the monograph of Butkovskiy (1969).

Example 5.3.1. An Optimal Pursuit Problem

Typical of boundary value problems arising from distributed parameter systems is the following well studied case (see, e.g., Axelband, 1967). Find the optimal control $f(t, x)$ which minimizes the integral

$$0 = \int_0^T \int_0^1 [(v(x, t) - u(x, t))^2 + \lambda^2 f^2(x, t)] \, dx \, dt$$

for a given integrable function v, where the state vector u satisfies the diffusion equation

$$\partial u / \partial t = (\partial^2 u / \partial x^2) + f(x, t), \qquad (x, t) \in (0, 1) \times (0, t)$$
$$u(x, 0) = u_0(x), \qquad u(0, t) = u(1, t) = 0$$

(5.3.8)

The control f is unconstrained except for the requirement that $f(\cdot, t) \in L_2[0, 1]$ for $t \in [0, T]$ and $\int_0^1 f^2(t, x) \, dx \in L_1[0, T]$.

It is well known that the operator $A = \partial^2 / \partial x^2$ is the self-adjoint infinitesimal generator of a strongly continuous semigroup $\{V(t); t \geq 0\}$ on $L_2^0[0, 1]$ [see, e.g., Kato (1966), p. 491], where L_2^0 is the subspace of L_2 of absolutely continuous functions vanishing at 0 and 1 with L_2 derivatives. Hence the state equation may be interpreted as the evolution equation

$$u' = Au + f, \qquad u(0) = u_0$$

with values in $L_2^0[0, 1]$. If we denote the inner produce on $L_2^0[0, 1]$ by $\langle \, , \, \rangle$

then the cost functional can be written as

$$0 = \int_0^T \{\langle v - u, v - u \rangle + \lambda^2 \langle f, f \rangle\} \, dt$$

The problem is formally identical to the finite-dimensional linear regulator considered in Section 3.4. It can be shown that the optimal control is given by the expression

$$f = (1/\lambda^2)\mu$$

where μ is calculated from the two-point boundary value problem

$$
\begin{aligned}
u' &= Au + f, & u(0) &= u_0 \\
\mu' &= -A^*\mu + u - v, & \mu(T) &= 0
\end{aligned}
\tag{5.3.9}
$$

Again, we note that this result is identical to that of the finite-dimensional case. Furthermore, it was already stated that A is the generator of the semigroup $\{V(t), t \geq 0\}$, and it is readily verified that $-A$ generates the semigroup $\{W(t), t \leq 0\}$ defined by $W(t) = V(-t)$, $t \leq 0$. (For a justification of taking adjoints in the case of non-self-adjoint semigroups see N.3.3). Hence it follows that system (5.3.9) satisfies the hypotheses imposed on (5.3.1).

The boundary value problem (5.3.9) is a particular example of the general system (5.3.1). Its solution is given by

$$u(t) = U(t)\mu(t) + w(t)$$

where U, w, and μ are obtained from the differential equations

$$
\begin{aligned}
U' &= (1/\lambda^2)I + AU + UA^* - U^2, & U(0) &= 0 \\
w' &= [A - U(t)]w + U(t)v(t), & w(0) &= 0 \\
\mu' &= [U(t) - A^*]\mu - v(t) + w(t), & \mu(T) &= 0
\end{aligned}
$$

It is straightforward to verify that the associated Volterra equations are

$$U(t) = \int_0^t V(t - r)[(1/\lambda^2)I - U^2(r)]V^*(t - r) \, dr$$

$$w(t) = S(t, 0)u_0 + \int_0^t S(t, r)U(r)v(r) \, dr$$

$$\mu(t) = - \int_t^T R(t, r)[w(r) - v(r)] \, dr$$

where the evolution operators S and R are solutions of the linear integral equations

$$S(t, r) = V(t - r) + \int_r^t V(t - s)U(s)S(s, r)\, ds, \qquad 0 \le r \le t$$

$$\text{(5.3.10)}$$

$$R(t, r) = V(r - t) + \int_r^t V(s - t)U(s)R(s, r)\, ds, \qquad 0 \le t \le r$$

Because $\| v(t) \|$ is integrable in t over $[0, T]$ the integrals of (5.3.11) are to be taken as Bochner integrals.

There exist several methods for establishing the existence of mild and strong solutions of Hermitian Riccati equations (Lions, 1971; Temam, 1971) and we shall state here that the above Riccati equation has a bounded mild solution on $[0, T]$. A proof of this result with the technique explored in Section 1.3 may be found in Section 6.4. As a result, Corollary 5.3.1 applies to the system (5.3.9) whose solution μ provides the optimal control $f = (1/\lambda^2)\mu$ for the diffusion equation (5.3.8).

In summary, we have shown that the invariant imbedding formation applies to the Hamiltonian system derived for the controlled diffusion equation so that the optimal control can be calculated from initial value problems. To conclude the discussion of this example let us comment on a possible numerical application. As in the Boltzmann transport problems, we can assume a concrete representation for $U(t)$. Suppose that U is written as the integral formula

$$(U(t)y)(\alpha) = \int_0^1 R(t, \alpha, \beta)y(\beta)\, d\beta$$

where y and R are assumed to be as smooth as required for subsequent operations. Substitution into the Riccati equation yields

$$\int_0^1 \frac{\partial R}{\partial t}(t, \alpha, \beta)y(\beta)\, d\beta$$
$$= \frac{1}{\lambda^2} \int_0^1 \delta(\alpha - \beta)y(\beta)\, d\beta + \int_0^1 \frac{\partial^2 R}{\partial \alpha^2}(t, \alpha, \beta)y(\beta)\, d\beta$$
$$+ \int_0^1 R(t, \alpha, \beta)(d^2y/d\beta)\, d\beta - \int_0^1 \int_0^1 A(t, \alpha, \gamma)R(t, \gamma, \beta)y(\beta)\, d\beta$$

This equation would have to hold for all differentiable y, hence the integrand can be isolated as:

$$\frac{\partial R}{\partial t}(t, \alpha, \beta) = \frac{1}{\lambda^2} \delta(\alpha - \beta) + \frac{\partial^2 R}{\partial \beta^2}(t, \alpha, \beta) + \frac{\partial^2 R}{\partial \beta^2}(t, \alpha, \beta)$$
$$- \int_0^1 R(t, \alpha, \gamma)R(t, \gamma, \beta)\, d\gamma, \qquad R(0, \alpha, \beta) = 0$$

Conversely, if a solution R of the integro–differential equation can be found then the corresponding operator $U(t)$ is a solution of the Riccati equation (N.3.4). ∎

Finally, it may be noted that the results of Section 3.4 on dynamic programming can be extended to abstract equations. In particular, this will allow an alternate derivation of several maximum principle presented in the literature (Meyer, 1970b). Similarly, multipoint boundary value for evolution problems can be handled. Indeed, the general conclusion is: If the given problem can be written formally as a boundary value problem for ordinary differential equations (in some sense), then the application of invariant imbedding becomes automatic. Its justification is a different matter.

NOTES

N.1.1. The concept of characteristics in an infinite-dimensional linear space is not new. Manninen (1960) considered the Cauchy problem $H(x, z'(x)) = 0$, $z = \eta(t)$ when $x = \mu(t)$ from a functional analytic point of view, where $H : D \subset R_x \times R_x' \rightarrow R^P$ is a twice differentiable operator defined on some domain D in the reflexive Banach space R_x. R_x' is the dual of R_x and R^P a p-dimensional linear space. From this operator H, Manninen derives the characters which for $p > 1$ are abstract total differential equations. The characteristics are then used to generate an integral surface through the given initial manifold. Our theory is more primitive but allows a more general setting and more general characteristic equations.

N.1.2. One may think here of the problems which arise when convergence of the discrete ordinate approximation to the continuous Boltzman equation is to be established (Nestell, 1967).

N.1.3. The invariant imbedding approach can be shown to be applicable to nonlinear evolution equation of the type discussed in Kato (1964). While not of known numerical value the method will provide existence theorems for certain partial differential equations with nonlinear source terms. The theory of nonlinear semigroups [see, e.g., Dorroh (1969)] may also provide an application where the mechanical association of a partial differential equation with its characteristic equations can be made rigorous. This topic has not been investigated.

N.2.1. There are some conceptual difficulties when relating partial differential equations to abstract evolution equations in L_p spaces since the solution of an abstract equation corresponds to an equivalence class of numerically valued functions. We refer to Hille and Phillips (1957, p. 68) for a discussion of how a solution of the partial differential equation can be recovered from the L_p solution.

N.3.1. The special case of time-dependent generators of the form $A + B(t)$ where A generates a semigroup and $B(t) \in L(X, X)$ is strongly continuous in t, is readily handled with the technique presented here and requires no additional tools for a theoretical foundation. For more general time dependent generators $A(t)$, corresponding, for example, to differential operators with nonconstant coefficients, our approximation approach still remains valid. Indeed, it is a requirement in the theory of evolution equations [see Kato (1964)] that for fixed t, $A(t)$ generates a semigroup $\{V(r)\}$. A promising approach to extending our theory would be to replace $A(t)$ by $A_h(t) = (V_t(h) - I)/h$ and consider the evolution operator generated by $A_h(t)$. A number of continuity theorems would be necessary for this approximation which do not appear to be available at this time. The construction of a parabolic evolution operator is presented in detail in Friedman (1969).

N.3.2. The level of complication to be expected when requiring the existence of strong solutions for the Riccati equation may become apparent from the local existence theorem given in Meyer (1970b) for parabolic evolution operators. Because of domain complications the problem likely is compounded for hyperbolic evolution operators.

N.3.3. We refer the reader to Hille and Phillips (1957, Section 14.2) for a discussion of the relation between semigroups, their generators and their adjoints. In the example the self-adjointness of A precludes all difficulties. For a more general control problem this simplification will be lost.

N.3.4. The question of finding an integral representation for the operator U is discussed in detail in Lions (1971). It is known that U is an invertible Hermitian operator so that the transformation $u(t) = U(t)\mu(t) + w(t)$ is a bona fide affine transformation on L_2. The Schwartz kernel theorem then assures that $U(t)$ can be represented by $U(t)\phi = \int_0^1 R(t, x, y)\phi(y)\, dy$ for an arbitrary test function ϕ. Isolating $R(t, x, y)$ in the differential equation for $U(t)$ leads to the above equation for R.

Chapter

6

Infinite-Dimensional Riccati Equations

6.1. LOCAL EXISTENCE THEOREMS

It is apparent from Chapter 5 that initial value methods for linear infinite-dimensional boundary value problems depend essentially on the
existence of solutions for the associated Riccati equations. This chapter
is intended to establish some sufficient conditions for the existence of
mild local and global solutions of usually non-Hermitian Riccati equations.
As was shown in Section 5.3 the restriction to mild solutions does not lead
to any loss of generality when solving boundary value problems.

As before we shall consider the Riccati equation

$$U' = B(t) + A(t)U - UD(t) - UC(t)U, \qquad U(0) = f \qquad (6.1.1)$$

where for fixed t the operator $U(t)$ belongs to the Banach space $L(X, Y)$
of bounded linear operators from the Banach space X to the Banach space Y.
The coefficients in (6.1.1) are assumed to be bounded linear operators of
"consistent dimensions," i.e., for fixed t, $B(t) \in L(X, Y)$, $A(t) \in L(Y, Y)$,
$D(t) \in L(X, X)$ and $C(t) \in L(Y, X)$. The continuity properties of the coefficients with respect to t determine whether Eq. (6.1.1) has a solution. The
simplest case is given if all coefficients are uniformly continuous in t. In
this case Theorems 5.1.1 and 5.1.3 apply. Specialized to Eq. (6.1.1) we can
rephrase these results as

THEOREM 6.1.1. Suppose that the coefficients of (6.1.1) are uniformly continuous in t on $[0, T]$. Then the Riccati equation has a unique differentiable solution $U(t)$ near the origin. Moreover, if $\| A(t) \| \leq a$, $\| B(t) \| \leq b$, $\| C(t) \| \leq c$, $\| D(t) \| \leq d$ and $\| f \| \leq i$ then the local solution can be continued over compact subsets of $[0, \hat{t})$, where $\hat{t}$ is given by Eqs. (1.1.5) or (1.1.8).

The proof of this theorem can be based on generating $U(t)$ from the characteristics, as was shown in earlier sections, or on establishing a differentiable solution for the Volterra equation

$$U(t) = f + \int_0^t F(r, U(r))\, dr \qquad (6.1.2)$$

Here $F(t, U(t))$ is meant to denote the right-hand side of Eq. (6.1.1) while the integral is the abstract Riemann integral in $L(X, Y)$. As is well known such a solution can be found from the Picard iteration

$$U_{n+1}(t) = f + \int_0^t F(r, U_n(r))\, dr, \qquad U_0(t) = f$$

It was observed before that the representation (6.1.2) lends itself to weakening the smoothness requirements on U and F. For example, under suitable hypotheses on F the iteration may be interpreted as taking place in the space of Bochner integrable functions from $[0, T]$ to $L(X, Y)$. Lipschitz continuity in U will again yield a local solution of (6.1.2) whose strong derivative is equal a.e. to $F(t, U)$. The next theorem makes this statement precise.

THEOREM 6.1.2. Suppose that on the rectangle

$$D = \{t \colon |t| \leq \alpha\} \times \{U \colon \| U - f \| \leq \beta\}$$

the function $F(t, U)$ is Bochner integrable in t for fixed U, Lipschitz continuous in U for fixed t and bounded such that $\| F(t, U) \| \leq m(t)$, where m is Lebesgue integrable on $[0, \alpha]$ uniformly on D. Then the Volterra equation (6.1.2) has a unique local solution near $(0, f) \in D$ whose derivative satisfies (6.1.1) a.e.

Proof. Let $\| F(t, U) - F(t, V) \| \leq K \| U - V \|$ for all t, U, $V \in D$ and let δ be so small that $\int_0^\delta m(t)\, dt \leq \beta$, $K\delta < 1$. Let $B(I, L)$ be the Banach space of Bochner integrable functions from $[0, \delta]$ to $L(X, Y)$ with norm $||| \cdot |||$. Then (6.1.2) maps $B(I, L)$ into itself and

$$||| U - V ||| \equiv \int_0^\delta \| U(t) - V(t) \| \, dt$$

$$= \int_0^\delta \left\| \int_0^t [F(r, U(r)) - F(r, V(r))] \, dr \right\| dt$$

$$\leq K\delta \int_0^\delta \| U(r) - V(r) \| \, dr$$

$$= K\delta \, ||| U - V |||$$

Hence by the contraction mapping theorem Eq. (6.1.2) has a unique solution in $B(I, L)$. The differentiability a.e. follows from the fact that $F(t, U(t)) \in B(I, L)$ [see Hille and Phillips (1957), p. 88]. ∎

Since strongly measurable functions with bounded and measurable norm over compact sets are Bochner integrable we obtain from Theorem 6.1.2 the following specific result.

COROLLARY 6.1.1. The Riccati equation has a unique continuous solution whenever the coefficients are strongly measurable and bounded and measurable in norm.

A somewhat different situation arises when the operator valued coefficients of (6.1.1) are strongly continuous in t. For example, they may be strongly continuous semigroups. It is known [see Hille and Phillips (1957), p. 85] that the above Picard iteration defines a sequence of operators in $L(X, Y)$. Since strongly continuous operators on compact intervals are also uniformly bounded it follows that the iterates are uniformly bounded and convergent to a uniformly continuous solution $U(t)$ of Eq. (6.1.1). Since also $F(t, U(t))x$ is continuous for all $x \in X$ we have

COROLLARY 6.1.2. The Riccati equation (6.1.1) has a unique local solution whenever the coefficients are strongly continuous in t. Moreover, $U(t)x$ is differentiable for all $x \in X$.

Finally we observe from the Volterra equation that the local solution can be continued in time as long as it stays bounded. Hence global solutions will exist provided the local solutions grow at most exponentially.

6.2. NON-HERMITIAN EQUATIONS WITH BOUNDED COEFFICIENTS

The continuity conditions and the Banach space setting, while sufficient for the existence of local solutions, appear to be too weak to establish bounded global solutions for Eq. (6.1.1). Henceforth, we shall restrict ourselves to linear operators between Hilbert spaces X and Y. We shall also need the following smoothness condition for the Riccati equation.

The coefficients of Eq. (6.1.1) have (uniformly) bounded weak derivatives [Kato (1966), p. 152]. For example, $\phi(t) = \langle B(t)x, y \rangle$ has a (uniformly) bounded derivative for all unit vectors x and y.

It should be remarked that the Hilbert space setting is restrictive for many boundary value problems where spaces of continuous functions or merely integrable functions provide a more natural setting than Hilbert function spaces. The additional continuity condition is not nearly as restrictive because in many applications the coefficients are either time independent or can be approximated by differentiable functions. It also may be noted from the inequality

$$\| B(t)x - B(\tau)x \|^2 = \langle B(t)x - B(\tau)x, B(t)x - B(\tau)x \rangle$$
$$= \langle B(t)x - B(\tau)x, v(\xi, x) \rangle (t - \tau)$$
$$\leq \| B(t)x - B(\tau)x \| \, \| v(\xi, x) \| \, | (t - \tau) |$$

where $v(\xi, x)$ is the weak derivative of $B(t)x$ evaluated at some $\xi \in (\tau, t)$, that the existence of bounded weak derivatives assures the strong continuity of the coefficients so that the Riccati equation has a local solution by Corollary 6.1.2.

As in Section 1.3 we shall need a Hermitian form associated with the Riccati equation (6.1.1) which is defined here as

$$\Omega(t, x, y) = \mathrm{Re}\{\langle B(t)x, y \rangle - \langle C(t)y, x \rangle + \langle A(t)y, y \rangle$$
$$- \langle D(t)x, x \rangle\} \tag{6.2.1}$$

where x and y are arbitrary unit vectors in X and Y, and where for ease of notation $\langle , \rangle$ stands for the inner product in both X and Y. We now can state the analog of Theorem 1.3.2 for Hilbert-space valued Riccati equations.

THEOREM 6.2.1. Assume that the coefficients of the Riccati equation (6.1.1) have uniformly bounded weak derivatives and that the initial value satisfies $\| f \| < 1$. Suppose further that there exists a constant $C > 0$ such that $\Omega(t, x, y) \leq -C$ for all $t \in [0, T]$ and all unit vectors $x \in X$ and $y \in Y$. Then the Riccati equation (6.1.1) has a unique solution on $[0, T]$.

Proof. We shall prove that $\| U(t) \| \leq 1$ on $[0, T]$. We know that $U(t)$ exists locally and that it is continuous in the uniform topology; hence $\| U(t) \|$ is a continuous function of t. Let us suppose that $\| U(t) \| < 1$ on $[0, t_0)$ and $\| U(t_0) \| = 1$ for some $t_0 \leq T$. Since

$$\| U(t_0) \| = \sup_{\substack{\|x\|=1 \\ \|y\|=1}} | \langle U(t_0)x, y \rangle |$$

it follows that there exist unit vectors $\hat{x}$ and $\hat{y}$ such that $\langle U(t_0)\hat{x}, \hat{y} \rangle \geq 1 - \varepsilon$ for arbitrary $\varepsilon > 0$. Let us write

$$U(t_0)\hat{x} = \alpha\hat{y} + \beta\hat{z} \qquad \text{and} \qquad U^*(t_0)\hat{y} = \gamma\hat{x} + \delta\hat{w}$$

where U^* is the adjoint of U, where $\alpha\hat{y}$ and $\gamma\hat{x}$ are the projections of $U(t_0)\hat{x}$ and $U^*(t_0)\hat{y}$ onto $\hat{y}$ and $\hat{x}$, and where $\beta\hat{z}$ and $\delta\hat{w}$ are the orthogonal complements with $\| \hat{z} \| = \| \hat{w} \| = 1$. Since $\| U(t_0)\hat{x} \| \leq 1$ and

$$\langle U(t_0)\hat{x}, \hat{y} \rangle = \langle \hat{x}, U^*(t_0)\hat{y} \rangle \geq 1 - \varepsilon$$

simple computation shows that $1 - \varepsilon \leq \alpha, \gamma \leq 1, | \beta |, | \delta | \leq 2\varepsilon$. Let us consider next the function $h(t) = \mathrm{Re}\langle U(t)\hat{x}, \hat{y} \rangle$. The assumptions imply that $| h(t) | < 1$ on $[0, t_0)$. We verify by differentiation and substitution for $U(t_0)\hat{x}$ and $U(t_0)\hat{y}$ that there exists a constant K_1 depending only on the bounds for the coefficients of (6.1.1) but not on $\varepsilon, \hat{x}, \hat{y}$ such that $h'(t_0) \leq \Omega(t, \hat{x}, \hat{y}) + K_1\varepsilon$. Hence for sufficiently small ε it is seen that $h'(t_0) < 0$. In addition, since the coefficients of (6.1.1) have bounded weak derivatives there exists a constant K_2 such that $| h''(t) | \leq K_2$ on $[0, t_0]$. We observe that for $t < t_0$ we can write

$$h(t) = h(t_0) + h'(t_0)(t - t_0) + (h''(\tilde{t})/2)(t - t_0)^2$$

where $t < \tilde{t} < t_0$. Using the estimate $\Omega(t_0, \hat{x}, \hat{y}) \leq -C$ we find that

$$h(t) \geq 1 - \varepsilon + (C - K_1\varepsilon)(t_0 - t) - K_2(t_0 - t)^2/2$$

The right-hand side of this expression assumes its maximum at

$$(t_0 - \hat{t}) = (C - K_1\varepsilon)/K_2$$

so that

$$h(\hat{t}) \geq 1 - \varepsilon + (C - K_1\varepsilon)^2/(2K_2)$$

Since C, K_1, and K_2 are independent of ε it follows that $h(\hat{t}) > 1$ for sufficiently small ε which contradicts $\| U(t) \| \leq 1$ on $[0, t_0)$. Thus $\| U(t) \| \leq 1$ on $[0, T]$ and the proof is complete. ∎

Two observations may be made here. If X and Y are finite dimensional then

$$\sup | \langle U(t_0)x, y \rangle | = \max | \langle U(t_0)x, y \rangle |$$

so that we may choose $\varepsilon = 0$. In this case

$$h'(t_0) = \Omega(t_0, \hat{x}, \hat{y}) < 0$$

is sufficient to conclude that $h(t) > 1$ for some $t < t_0$ regardless of the behavior of h''. Thus Theorem 6.2.1 reduces to Theorem 1.3.2. The second observation is that we may restrict the search for the unit vectors $\hat{x}$ and $\hat{y}$ to dense subsets of the infinite-dimensional unit spheres in the Hilbert spaces X and Y. The proof of Theorem 6.2.1 will remain unchanged while verification of the hypotheses may be made easier.

While Theorem 6.2.1 is a global result it does not reflect the existence of bounded local solutions since the Hermitian form $\Omega(0, x, y)$ may be positive or $\| f \|$ may exceed 1. A simple exponential transformation and an application of Theorem 6.2.1 allow us to overcome this defect. To this end let us define the operator valued function $V: [0, T] \rightarrow L(X, Y)$ by

$$V(t) = e^{-mt-k}U(t)$$

where m and k are positive constants, and where k is chosen such that $\| V(0) \| = e^{-k} \| f \| < 1$; the constant m remains to be determined. We verify by differentiation that V satisfies the Riccati equation

$$V' = e^{-mt-k}B(t) + A(t) - mIV - VD(t) - e^{mt+k}VC(t)V, \qquad V(0) = e^{-k}f.$$
$$(6.2.2)$$

Since the existence of V implies the existence of U we may rephrase Theorem 6.2.1 as follows.

THEOREM 6.2.2. Assume that the coefficients of (6.1.1) have uniformly bounded weak derivatives. Then there will exist a bounded solution $U(t)$ whenever constants m and k can be found such that the Hermitian form $\Omega(t, x, y, m, k)$ defined by

$$\Omega(t, x, y, m, k) = \text{Re}\{e^{-mt-k}\langle B(t)x, y \rangle + \langle A(t) - mIy, y \rangle$$
$$- \langle D(t)x, x \rangle - e^{mt+k}\langle C(t)y, x \rangle\} \qquad (6.2.3)$$

satisfies $\Omega(t, x, y, m, k) \leq -C < 0$ on $[0, T]$ for all vectors x and y in a dense subset of the unit spheres in X and Y, and such that $e^{-k} \| f \| < 1$.

We observe that for sufficiently large m we can force $\Omega(0, x, y, m, k)$ negative so that there will always be a local solution $U(t)$ near $t = 0$.

Example 6.2.1. Radiative Transfer in an Isotropic Slab (Continued)

Let us conclude this section by applying the above results to the Riccati equation (5.2.10) obtained in Example 5.2.1 for radiative transfer in a slab. It was shown that the equation has the form

$$U' = B + AU + UA = UBU, \quad U(0) = 0$$

where the operators A and B were formally defined by

$$Bh(\mu) = (c/2\mu) \int_\varepsilon^1 h(\lambda) \, d\lambda$$

$$Ah(\mu) = Bh(\mu) - (\sigma/\mu)h(\mu)$$

for an arbitrary integrable function h. In order to apply Theorem 6.2.1 the Riccati equation must be interpreted in a Hilbert space setting. It will be convenient to choose for both X and Y the space of all real square integrable functions on $[\varepsilon, 1]$ with respect to the Lebesgue–Stieltjes measure $\mu \, d\mu$. It is seen that for $\varepsilon > 0$ this space is equivalent to $L_2[\varepsilon, 1]$. Let us evaluate $\Omega(x, y)$. We find that

$$\mathrm{Re}\langle Bx, y \rangle = \langle Bx, y \rangle = \langle By, x \rangle = (c/2) \int_\varepsilon^1 x(\lambda) \, d\lambda \int_\varepsilon^1 y(\lambda) \, dd\lambda$$

$$\mathrm{Re}\langle Ax, x \rangle = \langle Bx, x \rangle - \sigma \int_\varepsilon^1 x^2(\lambda) \, d\lambda.$$

Thus

$$\Omega(x, y) = c \int_\varepsilon^1 x(\lambda) \, d\lambda \int_\varepsilon^1 y(\lambda) \, d\lambda + (c/2)\left(\int_\varepsilon^1 x(\lambda) \, d\lambda \right)^2$$

$$+ (c/2)\left(\int_\varepsilon^1 y(\lambda) \, d\lambda \right)^2 - \sigma \int_\varepsilon^1 x^2(\lambda) \, d\lambda - \sigma \int_\varepsilon^1 y^2(\lambda) \, d\lambda.$$

Since

$$c \int_\varepsilon^1 x(\lambda) \, d\lambda \int_\varepsilon^1 y(\lambda) \, d\lambda \leq (c/2)\left(\int_\varepsilon^1 x(\lambda) \, d\lambda \right)^2 + (c/2)\left(\int_\varepsilon^1 y(\lambda) \, d\lambda \right)^2$$

and

$$\left(\int_\varepsilon^1 x(\lambda) \, d\lambda \right)^2 \leq \int_\varepsilon^1 x^2(\lambda) \, d\lambda \int_\varepsilon^1 d\lambda$$

we obtain

$$\Omega(x, y) \leq [c(1 - \varepsilon) - \sigma]\left[\int_{\varepsilon}^{1} x^2(\lambda)\, d\lambda + \int_{\varepsilon}^{1} y^2(\lambda)\, d\lambda\right].$$

Thus for arbitrary $\varepsilon > 0$ it follows that the Riccati equation has a solution whenever $c \leq \sigma$. It may be observed that the extensive calculations of Bellman *et al.* (1963) of reflection functions for radiative transfer in a slab were carried out precisely over the range $0 < c \leq \sigma$. ∎

6.3. NON-HERMITIAN EQUATIONS WITH UNBOUNDED COEFFICIENTS

The approximation technique of Section 5.3 may be used to show that the results of the preceding sections also apply when A and D are closed linear operators. The key observation is that in the Hermitian form Ω the bilinear forms $\text{Re}\langle Ay, y\rangle$ and $\text{Re}\langle Dx, x\rangle$ need only be bounded above and below, respectively.

Specifically, this section will be concerned with the non-Hermitian Riccati equation

$$U' = B(t) + AU - UD - UC(t)U, \qquad U(0) = f \qquad (6.3.1)$$

where as before U is a linear operator from X into Y and where the coefficients are linear operators between the appropriate spaces. The earlier hypothesis of uniformly bounded weak derivatives is retained for the operators B and C, while for A and $-D$ we make the assumption already introduced in Section 5.3, namely that they are the infinitesimal generators of strongly continuous semigroups $\{V(t)\}$ and $\{W(t)\}$ on Y and X, respectively. As before, we denote by A_h and D_h for positive scalar h the bounded linear operators

$$A_h = (V(h) - I)/h, \qquad -D_h = (W(h) - I)/h$$

We also recall from Section 5.3 that the solution $U_h(t)$ of the Riccati equation

$$U_h' = B(t) + A_h U_h - U_h D_h - U_h C(t) U_h, \qquad U_h(0) = f \qquad (6.3.2)$$

converges strongly to the solution $U(t)$ of (6.3.1) as $h \to 0$ provided both exist. We have shown in Lemma 5.3.3 that (6.3.1) has a unique local solution. Thus in order to establish the boundedness of $U(t)$ on $[0, T]$ it is sufficient to establish the existence of a bound on $\|U_h(t)\|$ on $[0, T]$ which holds uniformly in h. In view of Theorem 6.2.2 we can state

THEOREM 6.3.1. Suppose that $e^{-k} \| f \| < 1$ and that the Hermitian form $\Omega_h(t, x, y, m, k)$ given by Eq. (6.2.3) for the Riccati equation (6.3.2) satisfies

$$\Omega_h(t, x, y, m, k) \leq -C < 0, \qquad t \in [0, T]$$

for arbitrary but fixed h, for constants m and k, and for all x and y in a dense subset of the unit spheres of X and Y. Then the Riccati equation (6.3.1) has a unique bounded solution on $[0, T]$.

Proof. For fixed $h > 0$ the operators A_h and D_h are uniformly bounded and similarly the second derivative of U is weakly bounded. Hence $\Omega_h(t, x, y, m, k) \leq -C$ and Theorem 6.2.2 assure the existence of $U_h(t)$ over $[0, T]$ such that $\| U_h(t) \| \leq e^{mt+k}$. Since this bound is independent of h it follows that also $\| U(t) \| \leq e^{mt+k}$. ∎

Example 6.3.1. Time-Dependent One-Dimensional Transport (Continued)

Let us apply Theorem 6.3.1 to Eq. (5.2.14) for time-dependent transport in a rod. Analogously to the system (5.2.16) these equations may be written formally as

$$u'(z) = -Au + \sigma(F - 1)u + \sigma Bv \qquad u(0) = a$$
$$v'(z) = -\sigma Bu + Av - \sigma(F - 1)v \qquad v(Z) = b \tag{6.3.1}$$

where $A = (1/c)\partial/\partial t$, where σ, B, and F are constants, and where a and b are given functions of time. We also shall assume that $c = 1$; this can always be achieved through the change of variable $\tau = ct$. The Riccati equation associated with (6.3.1) is seen to be

$$U' = \sigma BI + [\sigma(F - 1)I - A]U + U[\sigma(F - 1)I - A] + \sigma BU^2, \qquad U(0) = 0 \tag{6.3.2}$$

As the Hilbert spaces X and Y we shall choose the space $L_2(-\infty, \infty)$ of square integrable functions on $(-\infty, \infty)$. It is well known that A is the infinitesimal generator of the translation group $\{V(t)\}$ on (∞, ∞). If the initial values a and b of (6.3.1) belong to $L_2(-\infty, \infty)$ then it follows from Section 5.3 that the boundary value problem has a unique mild solution whenever the Riccati equation (6.3.2) has a solution over $[0, Z]$.

In order to use Theorem 6.3.1 we shall approximate A to be the central difference operator

$$A_h = [V(h) - V(-h)]/2h$$

It is well known that $\lim_{h \to 0} A_h x = Ax$ for all absolutely continuous functions $x \in L_2(-\infty, \infty)$ with square-integrable derivatives. Let us consider then the Riccati equation

$$U_h' = \sigma BI + [\sigma(F-1) - A_h]U_h$$
$$+ U_h[\sigma(F-1) - A_h] + \sigma BU_h^2, \qquad U_h(0) = 0$$

The Hermitian form for this equation is

$$\Omega_h(x, y) = 2\sigma B\langle x, y\rangle - \langle A_h x, x\rangle - \langle A_h y, y\rangle + \sigma(F-1)\}\{\langle x, x\rangle + \langle y, y\rangle\}$$

It is well known that the functions $\{t^n \exp(-t^2/2)\}$ are dense in $L_2(-\infty, \infty)$. Hence we may choose for the dense subset of the unit sphere the functions $\{\phi_n\}$ where $\phi_n = c_n t^n \exp(-t^2/2)$ and where c_n is chosen such that $\| \phi_n \| = 1$. We make the observation that by definition of A_h

$$\langle A_h\phi_n, \phi_n\rangle = \frac{c_n^2}{2h} \int_{-\infty}^{\infty} [(t+h)^n \exp(-(t+h)^2/2)$$
$$-(t-h)^n \exp(-(t-h)^2/2)t^n \exp(-t^2/2)] \, dt$$
$$= \frac{c_n^2}{2h} \left\{ \int_{-\infty}^{\infty} (t+h)^n \exp(-(t+h)^2/2)t^n \exp(-t^2/2) \, dt \right.$$
$$\left. - \int_{-\infty}^{\infty} s^n \exp(-s^2/2)(s+h)^n \exp(-(s+h)^2/2) \, ds \right\}$$

where $t - h = s$. Thus $\langle A_h\phi_n, \phi_n\rangle = 0$ for arbitrary h and n and the Hermitian form can be bounded as follows

$$\Omega_h(x, y) \leq 2\sigma(B + F - 1).$$

It is seen from Theorem 6.2.1 that U_h exists over $[0, Z]$ and remains bounded in norm by unity whenever $B + F < 1$. By Theorem 6.3.1 then U will exist over $[0, Z]$ and the system (6.3.1) is subcritical. It is well known [see Wing (1962), p. 11] that $B + F < 1$ is also a necessary condition for sub-critical transport in the steady state case ($c = \infty$). The result of Bailey (1964) for $B + F = 2$ is not obtainable by our method. We also would like to point out that backscatter at $z = 0$ does not change the above criterion for subcritical systems as long as the albedo is less than unity. ∎

On occasion enough information about A and $\{V(t)\}$ is known to allow us to conclude on theoretical grounds (without laborious computation) that $\langle A_h x, x\rangle$ is uniformly bounded above. One such criterion is given by the following lemma.

LEMMA 6.3.1. Let A be the generator of a semigroup $\{V(t)\}$ with $\| V(t) \| \leq e^{\omega t}$. Let $A_h = (V(h) - I)/h$ for a scalar h. Then

$$\mathrm{Re}\langle A_h x, x \rangle \leq \omega \langle x, x \rangle$$

for all $x \in X$ uniformly in h.

Proof. For arbitrary $x \in X$ define $z = x - A_h x/\lambda$ where λ is a positive number. With this definition we have

$$\| x + x/\lambda h \| = \| z + V(h)x/\lambda h \| \leq \| z \| + e^{\omega h} \| x \|/\lambda h$$
$$(1 + (1 - e^{\omega h})/\lambda h) \| x \| \in \| z \| = \| I - A_h x/\lambda \|.$$

Multiplying by λ and estimating $1 - e^{\omega h} \geq -\omega h$ we obtain

$$(\lambda - \omega) \| x \| \leq \| (\lambda I - A_h)x \|.$$

Squaring both sides, using inner product notation and letting $\lambda \to \infty$ we obtain the final result

$$\mathrm{Re}\langle A_h x, x \rangle \leq \omega \langle x, x \rangle. \quad \blacksquare$$

Note that in the above example of transport in a rod the translation group $\{V(t)\}$ satisfies $\| V(t) \| = 1$ so that $\omega = 0$. Hence it follows immediately that $\langle A_h x, x \rangle \leq 0$ which in turn is sufficient to establish the existence of a bounded mild solution of the abstract Riccati equation. However, in most applications only the operator A is given while the semigroup $\{V(t)\}$ may not be known. Thus a characterization of those operators is needed which give rise to a semigroup $\{V(t)\}$ with $\| V(t) \| \leq e^{\omega t}$. An easily computed criterion is given by the following converse of Lemma 6.3.1.

LEMMA 6.3.2. Let A be a densely defined closed linear operator on X and suppose that $\mathrm{Re}\langle Ax, x \rangle \leq \omega \langle x, x \rangle$ for all $x \in D(A)$ and some real constant ω. Then A is the generator of a semigroup $\{V(t)\}$ with $\| V(t) \| \leq e^{\omega t}$.

Proof. This result is well known and may be proved as follows. For arbitrary $x \in D(A)$ and real λ we have

$$\langle (\lambda I - A)x, (\lambda I - A)x \rangle = \lambda^2 \langle x, x \rangle - 2\lambda \, \mathrm{Re}\langle Ax, x \rangle + \langle Ax, Ax \rangle \geq 0$$

The minimum of this expression is assumed at $\lambda = \mathrm{Re}\langle Ax, x \rangle/\langle x, x \rangle$ so that the preceding inequality requires that $\langle Ax, Ax \rangle \geq (\mathrm{Re}\langle Ax, x \rangle)^2/\langle x, x \rangle$. Hence

$$\| (\lambda I - A)x \|^2 \geq (\lambda \| x \| - \mathrm{Re}\langle Ax, x \rangle/\langle x, x \rangle)^2$$

or

$$\| (\lambda I - A)x \| \geq (\lambda - \omega) \| x \| \quad \text{for} \quad \lambda > \omega$$

By the Hille–Yosida–Phillips theorem A is the generator of a semigroup $\{V(t)\}$ with $\| V(t) \| \leq e^{\omega t}$. $\blacksquare$

Note that for transport in a rod we find

$$\langle Ax, x \rangle = \int_{-\infty}^{\infty} x'(t)x(t) \, dt = x^2(t)/2 \Big|_{-\infty}^{\infty} = 0$$

for all absolutely continuous functions in L_2 with L_2 derivatives.

An extension of these results to semigroups $\{V(t)\}$ with $\| V(t) \| \leq Me^{\omega t}$, $M > 1$ has not yet been found.

6.4. HERMITIAN EQUATIONS

We have seen in Examples 2.10.1 and 5.3.1 that Hermitian Riccati equations arise naturally in the theory of unconstrained optimal control for linear state and quadratic cost equations. Consequently, finite- and infinite-dimensional Hermitian equations have received considerable attention (see Reid, 1972; Lions, 1971). We shall indicate here how the theorems of the last section can be specialized to rederive some of the known results for Hermitian equations. Let us consider first equations with bounded coefficients of the form

$$U' = B(t) + A(t)U + UA^*(t) - UC(t)U, \quad U(0) = f \quad (6.4.1)$$

where as before all coefficients have bounded weak derivatives. In addition it is assumed that B and C are self-adjoint while A^* denotes the adjoint of A. It is readily verified that U^* and U satisfy Eq. (6.4.1) so that by Corollary 6.1.2 they coincide. Thus the local solution of the Hermitian–Riccati equation is self-adjoint. This observation allows us to rephrase Theorem 6.2.2 as follows.

THEOREM 6.4.1. Assume that the coefficients of (6.4.1) have uniformly bounded weak derivatives. Suppose further that there exist constants $C > 0$ and m, k such that $e^{-k} \| f \| < 1$ and

$$\Omega(t, x, m, k) \leq | \langle [e^{-mt-k}B(t) - e^{mt+k}C(t)]x, x \rangle |$$
$$+ 2 \operatorname{Re} \langle A(t)x, x \rangle - m \leq -C \quad (6.4.2)$$

for $t \in [0, T]$ and for all unit vectors x in a dense subset of the unit sphere of the Hilbert space X. Then the Riccati equation (6.4.1) has a unique global solution on $[0, T]$.

The proof of the theorem is identical to that of Theorem 6.2.2 provided we use the representation $\| U(t) = \sup_{\|x\|=1} | \langle U(t)x, x \rangle |$ instead of $\sup_{\|x\|=\|y\|=1} | \langle U(t)x, y \rangle |$. As a consequence, $\hat{y}$ in the earlier proof may be replaced by $+\hat{x}$ or $-\hat{x}$. Substitution of $\pm\hat{x}$ into the Hermitian form (6.2.3) leads to (6.4.2).

A special case is given if $B(t)$ and f are nonnegative definite. It follows from the (dimensionless) Lemma 1.3.1 and Corollary 1.3.1 that $\langle U(t)x, x \rangle \geq 0$ on $[0, T]$. This allows us to delete the absolute values in (6.4.2) so that the hypothesis of Theorem 6.4.1 becomes

$$\Omega(t, x, m, k) \equiv \langle [e^{-mt-k}B(t) - e^{mt+k}C(t)]x, x \rangle$$
$$+ 2\,\mathrm{Re}\langle A(t)x, x \rangle - m \leq -C \qquad (6.4.3)$$

If $C(t)$ is nonnegative then this form will always be negative for sufficiently large m. Hence we obtain from Theorem 6.4.1

COROLLARY 6.4.1. If f, B and C are nonnegative definite then the Riccati equation (6.4.1) has a unique solution on $[0, T]$.

It may be noted that implicit in our approach here is the requirement that the coefficients have bounded weak derivatives. If in a particular application this property is not present then the conclusion of Corollary 6.4.1 can be obtained from the (dimensionless) Corollary 1.3.2.

So far we have assumed that $A(t)$ is a uniformly continuous bounded linear operator. However, all results of this section remain valid when $A(t) = A$ and $A^*(t) = A$ are the closed infinitesimal generators of strongly continuous semigroups $\{V(t)\}$ and $\{W(t)\}$ such that $\langle A_h x, x \rangle$ and $\langle A_h^* x, x \rangle$ are bounded above uniformly with respect to h. For example, we can obtain the following analog of Corollary 6.4.1.

THEOREM 6.4.2. Suppose that f, B, and C are nonnegative definite, that A is the generator of the semigroup $\{V(t)\}$ with $\| V(t) \| \leq e^{\omega t}$, and that $D(A) \subset D(A^*)$. Then the Riccati equation (6.4.1) has a unique mild solution on $[0, T]$.

Proof. It is known [Hille and Phillips (1957), p. 421] that A^* generates the semigroup $\{V^*(t)\}$. Hence the mild solution $U(t)$ of the Riccati equation

exists locally and can be approximated by $U_h(t)$. The existence of a uniformly bounded $U_h(t)$ follows from Lemma 6.3.1 and Corollary 6.4.1. ∎

Example 6.4.1. An Optimal Pursuit Problem (Continued)

Let us conclude this section by considering again the Riccati equation
of Example 5.3.1 for the controlled diffusion equation. It is given by

$$U' = (1/\lambda^2)I + AU + UA - U^2, \qquad U(0) = 0$$

where A is the differential operator $\partial^2/\partial z^2$ on $L_2(0, 1)$ with domain $D(A)$
$= \{x: 0 = x(0) = x(1), x'' \in L_2(0, 1)\}$. It is well known and straightforward
to verify that A is self-adjoint, densely defined and closed. Moreover,
integration by parts shows that

$$\langle Ax, x \rangle \leq 0$$

for all $x \in D(A)$. Hence it is dissipative and by Lemma 6.3.2 the generator
of a strongly continuous semigroup $\{V(t)\}$ with $\| V(t) \| \leq 1$. Thus the
Riccati equation (6.4.4) has a unique mild solution by Theorem 6.4.2. ∎

References

Allen, R., and Wing, G. M. (1970). A numerical algorithm suggested by problems of transport in periodic media. *J. Math. Anal. Appl.* **29**, 141–157.

Allen, R., Wing, G. M., and Scott, M. (1969). Solution of a certain class of nonlinear two point boundary value problems. *J. Computational Phys.* **4**, 250–257.

Ames, W. (1965). "Nonlinear Partial Differential Equations in Engineering." Academic Press, New York.

Apostol, T. (1957). "Mathematical Analysis." Addison-Wesley, Reading, Massachusetts.

Atkinson, B., and Daoud, I. (1968). The analogy between microbiological "reactions" and heterogeneous catalysis. *Trans. Inst. Chem. Engrs.* **46**, 19–24.

Axelband, E. (1967). A solution to the optimal pursuit problem for distributed parameter systems. *J. Computational System Sci.* **1**, 261–286.

Babuska, I., Prager, M., and Vitasek, E. (1966). "Numerical Processes in Differential Equations." Wiley (Interscience), New York.

Bailey, P. (1964). A rigorous derivation of some invariant imbedding equations of transport theory. *J. Math. Anal. Appl.* **8**, 144–169.

Bailey, P., and Wing, G. (1965). Some recent developments in invariant imbedding with applications. *J. Mathematical Phys.* **6**, 453–462.

Bailey, P., Shampine, L., and Waltman, P. (1968). "Nonlinear Two-Point Boundary Value Problems." Academic Press, New York.

Belford, G. (1969). An initial value problem approach to the solution of eigenvalue problems. *SIAM J. Numer. Anal.* **6**, 99–103.

Bellman, R. (1953). "Stability Theory of Differential Equations." McGraw-Hill, New York.

Bellman, R. (1957). "Dynamic Programming." Princeton Univ. Press, Princeton, New Jersey.

Bellman, R., Kagiwada, H., Kalaba, R., and Prestrud, M. (1963). "Invariant Imbedding and Radiative Transfer in Slabs of Finite Thickness." Amer. Elsevier, New York.

Bellman, R., and Kalaba, R. (1964). Dynamic programming, invariant imbedding and quasi linearization: Comparisons and interconnections. *Rand Corp. Memor.* RM-4038-PR.

Bellman, R., Cooke, K., Kalaba, R., and Wing, G. M. (1965). Existence and uniqueness theorems in invariant imbedding. I. Conservation principles. *J. Math. Anal. Appl.* **10**, 234–244.

Bellman, R., and Cooke, K. (1965). Existence and uniqueness theorems in invariant imbedding. II. Convergence of a new difference algorithm. *J. Math. Anal. Appl.* **12**, 247–253.

Bellman, R. (1967). "Introduction to the Mathematical Theory of Control Processes." Academic Press, New York.

Berezin, I., and Zhidkov, N. (1965). "Computing Methods," Vol. II. Pergamon Press, Oxford.

Butkovskiy, A. (1969). "Theory of Optimal Control of Distributed Parameter Systems." Amer. Elsevier, New York.

Butzer, P., and Berens, H. (1967). "Semigroups of Operators and Approximation." Springer-Verlag, Berlin and New York.

Cannon, J., and Hill, C. D. (1970). On the movement of a chemical reaction interface. *J. Math. Mech.* **20**, 429–454.

Cannon, J., and Meyer, G. (1971). On diffusion in a fractured medium. *SIAM J. Appl. Math.* **20**, 434–448.

Cannon, J., Douglas, J., Jr., and Hill, C. D. (1967). A multiboundary Stefan problem and the disappearance of phases. *J. Math. Mech.* **17**, 21–34.

Ciarlet, P., Schultz, M., and Varga, R. (1967). Numerical methods of high order accuracy for nonlinear boundary value problems. I. One-dimensional problems. *Numer. Math.* **9**, 394–430.

Ciarlet, P., Schultz, M., and Varga, R. (1968). Numerical methods of high order accuracy for nonlinear boundary value problems. II. Nonlinear boundary conditions. *Numer. Math.* **11**, 331–345.

Ciment, M., and Guenther, R. (1969). Numerical solution of a free boundary value problem for parabolic equations. *Marathon Oil Co. Tech. Memo.* 69–22.

Coddington, E., and Levinson, N. (1955). "Theory of Ordinary Differential Equations." McGraw-Hill, New York.

Collatz, L. (1960). "The Numerical Treatment of Differential Equations." Springer-Verlag, Berlin and New York.

Courant, R., and Hilbert, D. (1962). "Partial Differential Equations." Wiley (Interscience), New York.

Courant, R., Isaacson, E., and Rees, M. (1952). On the solution of nonlinear hyperbolic differential equations by finite difference. *Comm. Pure Appl. Math.* **5**, 243–255.

Dieudonné, J. (1969). "Foundations of Modern Analysis." Academic Press, New York.

Dorroh, J. (1969). A nonlinear Hille-Yosida-Phillips theorem. *J. Functional Anal.* **3**, 345–353.

Dunford, N., and Schwartz, J. (1957). "Linear Operators." Wiley (Interscience), New York.

Elmas, M. (1970). The theory of fluidized bed coating. *Chem. Engrg.* **1**, 217–230.

Evans, H. (1968). "Laminar Boundary Layer Theory." Addison-Wesley, Reading, Massachusetts.

Falb, P., and Kleinman, D. (1966). Remarks on the infinite dimensional Riccati equation. *IEEE Trans. Automatic Control* **AC-11**, 534–536.

Forsythe, G., and Wasow, W. (1960). "Finite Difference Methods for Partial Differential Equations." Wiley, New York.

Friedman, A. (1967). Optimal control in Banach spaces. *J. Math. Anal. Appl.* **19**, 35–55.

Friedman, A. (1968). One-dimensional Stefan problem with nonmonotone free boundary. *Amer. Math. Soc. Transl.* **133**, 89–114.

Friedman, A. (1969). "Partial Differential Equations." Holt, New York.

Gelfand, I., and Fomin, S. (1963). "Calculus of Variations." Prentice-Hall, Englewood Cliffs, New Jersey.

Golberg, M. (1971). A generalized invariant imbedding equation. *J. Math. Anal. Appl.* **34**, 590–601.

Goodman, T., and Lance, G. (1956). The numerical solution of two-point boundary value problems. *Math. Comp.* **10**, 82–86.

Gourlay, A. (1970). A note on trapezoidal methods for the solution of initial value problems. *Math. Comp.* **24**, 629–633.

Guderley, K., and Nikolai, P. (1966). Reduction of two-point boundary value problems in a vector space to initial value problems by projections. *Numer. Math.* **8**, 270–289.

Hansen, A. (1964). "Similarity Analyses of Boundary Value Problems in Engineering." Prentice-Hall, Englewood Cliffs, New Jersey.

Hille, E., and Phillips, R. (1957). "Functional Analysis and Semigroups." Colloquium Publ. 31, Amer. Math. Soc., Providence, Rhode Island.

Jacobson, D. (1970). New conditions for boundedness of the solution of a matrix Riccati differential equation. *J. Differential Equations* **8**, 258–263.

Kamke, E. (1943). "Differentialgleichungen. Lösungsmethoden und Lösungen." Akad. Verlagsges., Leipzig.

Kamke, E. (1950). "Differentialgleichungen reeller Funktionen." 2nd. rev. ed. Akad. Verlagsges., Leipzig.

Kato, T. (1964). Nonlinear evolution equations in Banach spaces. *Proc. Symp. Appl. Math.* XVII, 50–67, Amer. Math. Soc., Providence, Rhode Island.

Kato, T. (1966). "Perturbation Theory for Linear Operators." Springer-Verlag, Berlin and New York.

Keller, H. (1968). "Numerical Methods for Two-Point Boundary Value Problems." Blaisdell, Waltham, Massachusetts.

Keller, H. (1971). Shooting and imbedding for two-point boundary value problems. *J. Math. Anal. Appl.* **36**, 598–610.

Kruzhov, S. (1967). On some problems with unknown boundaries for the heat conduction equation. *Prikl. Mat. Meh.* **31**, 1009–1020.

Lax, P. (1954). Weak solutions of nonlinear hyperbolic equations and their numerical computation. *Comm. Pure Appl. Math.* **7**, 84–87.

Lax, P. (1969). Nonlinear partial differential equations and computing. *SIAM Rev.* **11**, 9–19.

Lee, E. S. (1968). "Quasi Linearization and Invariant Imbedding." Academic Press, New York.

Lions, J. (1971). "Optimal Control of Systems Governed by Partial Differential Equations." Springer-Verlag, New York.

Manninen, J. (1960). Zur Charakteristikentheorie von Systemen partieller Differentialgleichungen erster Ordnung. *Ann. Acad. Sci. Fenn. Ser. AI* **33**, 3–37.

March, N. (1952). Thomas-Fermi fields for molecules with tetrahedral and octahedral symmetry. *Proc. Camb. Phil. Soc.* **48**, 665–675.

Mason, J. (1966). Approximate formulae for a variety of solutions of the Thomas-Fermi equation. *Univ. of Maryland Techn. Note* BN-430.

Meyer, G. (1968). On a general theory of characteristics and the method of invariant imbedding. *SIAM J. Appl. Math.* **16**, 488–509.

Meyer, G. (1970a). The invariant imbedding equations for multipoint boundary value problems. *SIAM J. Appl. Math.* **18**, 433–453.

Meyer, G. (1970b). On fixed time control problems in a Banach space. *SIAM J. Control* **8**, 383–395.

Meyer, G. (1970c). On a free interface problem for linear ordinary differential equations and the one-phase Stefan problem. *Numer. Math.* **16**, 248–267.

Meyer, G. (1971). A numerical method for two-phase Stefan problems. *SIAM J. Numer. Anal.* **8**, 555–568.

Meyer, G., Keller, H., and Couch, E., Jr. (1972). Thermal models for road airstrips and building foundations in permafrost regions. *J. Canad. Petrol. Tech.* **11**, 1–13.

Miele, A. (1970). Method of particular solutions for linear two-point boundary value problems. *J. Opt. Theory Appl.* **2**, 260–273.

Na, T., and Hansen, A. (1968). General group theoretic transformations from boundary value into initial value problems. *NASA rept.* CR-61218.

Nelson, P., and Altom, D. (1971). Hybrid solution of partial differential equations by application of invariant imbedding to the serial method. *Simulation* **17**, 147–154.

Nestell, M. (1967). The convergence of the discrete ordinates method for integral equations of anisotropic radiative transfer. Ph.D. Thesis, Oregon State University.

Ortega, J., and Rheinboldt, W. (1970). "Iterative Solution of Nonlinear Equations in Several Variables." Academic Press, New York.

Reid, W. (1972). "Riccati Differential Equations." Academic Press, New York.

Roberts, S., and Shipman, J. (1970). The method of adjoints and complementary functions in two-point boundary value problems. *Internat. J. Computer Math.* **2**, 269–277.

Sage, A. (1968). "Optimum Systems Control." Prentice-Hall, Englewood Cliffs, New Jersey.

Sackett, G. (1971). Numerical solution of a parabolic free boundary problem arising in statistical decision theory. *Math. Comp.* **25**, 425–434.

Sage, A., and Melsa, J. (1971). "System Identification." Academic Press, New York.

Scott, M., Shampine, L., and Wing, G. M. (1969). Invariant imbedding and the calculation of eigenvalues for Sturm-Liouville systems. *Computing* **4**, 10–23.

Scott, M. (1970). Numerical solution of unstable initial value problems by invariant imbedding. *Comput. J.* **13**, 397–400.

Scott, M. (1972). A bibliography on invariant imbedding and related topics. *Sandia Laboratories Tech. Note* SC-B-71 0886.

Segal, I. (1963). Nonlinear semigroups. *Ann. Math.* **78**, 339–364.

Shampine, L., and Thompson, R. (1970). Difference methods for nonlinear first-order hyperbolic systems of equations. *Math. Comp.* **24**, 45–56.

Sherman, B. (1971). General one-phase Stefan problems and free boundary problems for the heat equation with Cauchy data prescribed on the boundary. *SIAM J. Appl. Math.* **20**, 555–570.

Spingarn, K. (1971). A comparison of numerical methods for solving optimal control problems. *IEEE Trans. Vol. AES* **7**, 73–78.

Taufer, J. (1966). On factorization method. *Appl. Mat.* **11**, 427–457.

Temam, R. (1971). Sur l'equation de Riccati associée à des opérateurs non bornés, en dimension infinie. *J. Functional Analysis* **7**, 85–115.

Vainberg, M. (1964). "Variational Methods for the Study of Nonlinear Operators." Holden-Day. San Francisco.

Vishnevetsky, R. (1968). A new stable computing method for the serial hybrid computer integration of partial differential equations. *AFIPS Proc.* **32**, 143–150.

Weinel, E. (1965). Lineare Randwertaufgaben und Grassmannsche Koordinaten in projektiven Funktionenräumen. *Wiss. Z. Friedrich-Schiller-Univ. Jena/Thüringen* **14**, 409–416.

Wentzel, T. (also written Ventcel') (1960). A free boundary problem for the heat equation. *Dok. Akad. Nauk SSSR* **131**, 1000–1003.

Wing, G. M. (1962). "An Introduction to Transport Theory." Wiley, New York.

Zwas, G., and Abarbanel, S. (1971). Third and fourth order accurate schemes for hyperbolic equations of conservation law form. *Math. Comp.* **25**, 229–236.

Index